COSMIC
PARADOXES
Third Edition

COSMIC
PARADOXES
Third Edition

Julio A Gonzalo
Universidad Autónoma de Madrid, Spain
Universidad San Pablo CEU, Madrid, Spain

 **World Scientific**

NEW JERSEY · LONDON · SINGAPORE · BEIJING · SHANGHAI · HONG KONG · TAIPEI · CHENNAI · TOKYO

Published by

World Scientific Publishing Co. Pte. Ltd.
5 Toh Tuck Link, Singapore 596224
USA office: 27 Warren Street, Suite 401-402, Hackensack, NJ 07601
UK office: 57 Shelton Street, Covent Garden, London WC2H 9HE

Library of Congress Control Number: 2022030148

British Library Cataloguing-in-Publication Data
A catalogue record for this book is available from the British Library.

COSMIC PARADOXES
Third Edition

ISBN 978-981-126-206-7 (hardcover)
ISBN 978-981-126-309-5 (paperback)
ISBN 978-981-126-207-4 (ebook for institutions)
ISBN 978-981-126-208-1 (ebook for individuals)

For any available supplementary material, please visit
https://www.worldscientific.com/worldscibooks/10.1142/13018#t=suppl

Typeset by Stallion Press
Email: enquiries@stallionpress.com

Printed in Singapore

To my parents, and to Saint John Paul II

Foreword

Paradoxes — either real or apparent contradictions — have been part and parcel of cosmology for centuries. Many have tried to interpret the dark night paradox (Olber's paradox) in such a way as to save the presumed infinity of the universe with no success.

In this book cosmic paradoxes are seriously examined from an honest, conservative point of view. What does a conservative approach mean? It is an approach which does not impose *a priori* conditions (such as the necessary flatness of space-time all the way through cosmic expansion) disregarding the evidence to the contrary if there is any. And it is an approach which, on confronting any objective entity, ranging from an atom to a galaxy and even the whole cosmos, does not take lightly *energy conservation*. This was done in the 1950s by Gold, Bondi and Hoyle, in their *Steady State Theory*, and in the 1980's, up to now, by Guth, Linde *et al.*, in the *Inflationary Theory*. Of course, if the requisite of energy conservation is relaxed one may enjoy unlimited freedom. But this freedom remains suspicious even after receiving support (real or imaginary) from Quantum Mechanics.

The Big Bang model, anticipated by Georges Lemaitre and formulated some years later by Gamow, Alpher and Herman, is certainly one of the most spectacular scientific successes in the entire history of physics. It remains today, in my opinion, surrounded by too much speculation, without real physical support, for the time being.

Recently, at a scientific meeting in Philadelphia, one attending colleague dared to suggest that the TOE should be called the "Theory *of* Everything *except* Experimental Reality". As Planck once said: "Theorists are numerous and paper is patient". Because of the finiteness of the speed of light, here and now, we are seeing a *superposition* of snapshots, coming from a multitude of galaxies when they were at distances ranging from the relatively close Andromeda to the most distant protogalaxies, or quasars, emitted at times ranging from a few million light years to around ten billion light years. It is therefore absolutely mandatory to take this into consideration if we are to interpret correctly cosmological data.

Stanley Jaki in *The Road of Science and the Ways to God* recalled what Planck took as the paradox of relativity: instead of relativizing everything, it unfolded *absolute*, objective aspects of the physical world. In October 26–27, 1990, a Symposium on *Physics and Religion in Perspective* was held in Madrid, sponsored by BBV and UNED. The opening lecture, by Prof. Stanley L. Jaki (Princeton) was on "Physics and the Universe: From the Sumerians to the late twentieth century". Professor Jaki's lecture is reproduced in an appendix at the end of this book.

In August 16–20, 1993 a Summer Course on *Astrophysical Cosmology*, organized by the Universidad Complutense, Madrid, and co-sponsored by the Spanish Royal Academy of Sciences (Real Academia Española de Ciencias Exactas Físicas y Naturales), was held at the historic site of El Escorial, near Madrid. The Program of the course included talks by a few true luminaries of modern scientific cosmology: Ralph A. Alpher, George F. Smoot, John C. Mather, who acted as Co-Director of the Summer Course, aided by Jose Mª Torroja (Secretary of the Spanish Royal Academy of Science), other distinguished foreign speakers (Jerome Lejeune, Hans Elsaesser, Stanley L. Jaki) and a number of Spanish speakers from different institutions (Jose L. Sanchez Gomez, Antonio Fernandez Rañada, Manuel Catalan and Julio A. Gonzalo). The Summer Course was coordinated by Rodolfo Nuñez de las Cuevas, Ignacio Cantarell and Julio A. Gonzalo, as Secretary, and was whole heartedly supported from the beginning by Miguel Angel Alario, Dean of the Faculty of Chemistry (UCM) and General Scientific Coordinator of the El Escorial Courses that year.

The first draft of the Summer Course Program was sketched in June 1992, on the occasion of a conference on "Is there such a thing as a last word in Physics?", given by Stanley L. Jaki at the Spanish Royal Society, followed by an animated question and answer period. The confirmation of the Planck characteristic distribution of the cosmic background radiation spectrum and the detection of their minute anisotropies measured by NASA's COBE satellite completed in a satisfactory way the description of the initial moments of the universe's thermal history. Some of the protagonists of this story were there, at the El Escorial Summer Course. I think it is fitting to reproduce here in full the English version of the 1993 Summer Course at El Escorial.

16 Monday	10:00	R. A. Alpher (Schenectady, N. Y.): *"The Cosmic Background Radiation (CBR)"*
	11:30	G. F. Smoot (Berkeley): *"Satellite observations of the CBR anisotropy"*
	16:00	Round Table: *"The significance of the CBR"* Moderator: J. M. Torroja; R. A. Alpher, G. F. Smoot, J. C. Mather, H. Elsaesser.
17 Tuesday	10:00	J. L. Sánchez Gómez (Madrid) *"Cosmology and elementary particles"*
	11:30	A. Fernandez Rañada (Madrid) *"Thermal history of the universe"*
	16:00	Round Table: *"Evaluation of the big-bang model"* Moderator: A. Tiemblo; J. L. Sanchez Gómez, A. Fernandez Rañada, J. M. Quintana, J. A. Gonzalo.
18 Wednesday	10:00	J. C. Mather (NASA): *"The COBE project: achievements in perspective"*
	11:30	J. Lejeune (Paris): *"Neurophysiology of human intelligence"*

Brief Summaries of the lectures given by the invited speakers at the Summer Course are given below.

R. A. Alpher: "The Cosmic Background Radiation"

It is commonly accepted that the observable universe evolved from a high density and high temperature primitive state some fifteen to twenty billion years ago. The event ended up being called the "Big Bang". The most successful relativistic model for this universe is based upon the homogeneous and isotropic expansion of an ideal fluid from an initial state dominated by radiation to the present state dominated by matter. The model encompasses the thermonuclear synthesis of light elements during the first few minutes to the transition from an expansion controlled by radiation to other controlled by matter after a few hundred thousand years. In subsequent epochs the present structure of the universe developed and the heavier elements were formed in the stars.

G. F. Smoot: "Satellite observations of the CBR anisotropy"

The Cosmic Background Explorer (COBE) satellite was developed to measure the diffuse infrared and microwave radiation from the early universe, to the limits set by the astrophysical foregrounds. COBE has three instruments: The Differential Microwave Radiometer (DMR) maps the cosmic radiation precisely. The Far Infrared Spectrophotometer (FIRAS) compares the cosmic background radiation spectrum with that of a precise black-body. The Diffuse Infrared Background Experiment (DIRBE) searches the accumulated light of primeval galaxies and stars. FIRAS has measured the cosmic background radiation spectrum nearly 1000 times more precisely than previous observations. The DMR discovered the cosmic background to be anisotropic at a level of one part in 10^5. The results strongly constrain cosmological models and the formation of structure in the present universe. The results from COBE are in good quantitative agreement with the calculations of the hot Big Bang model. The DMR data are in agreement with inflationary model predictions.

J. L. Sanchez Gomez: "Cosmology and elementary particles"

Brief summary of the standard cosmology model: expansion of the universe, cosmic background radiation and primordial nucleosynthesis. Cosmological consequences of particle physics: matter-antimatter symmetry, the problem of the exotic dark matter and the curvature of the universe, phase transitions in the primordial universe and inflationary Big Bang model.

The aim of the talk is to present the last advances in elementary particle physics of interest for cosmology.

The talk will have a markedly informative and descriptive character. Excessively formal questions will be avoided and the fundamental aspects of the intriguing relationship between elementary particles and cosmology will be given, avoiding, as much as possible, technical language.

A. Fernandez Rañada: "Thermal history of the universe"

1. The universe's expansion and Hubble's law: A gas of galaxies; the recession of galaxies; the age of the universe.
2. The explosion: Thermal history of the universe. From $t = 10^{-11}$ s to today; from $t = 10^{-43}$ s (Planck's time) to $t = 10^{-11}$ s; from $t = 0$ to $t = 10^{-43}$ s.
3. Test and problems of the standard Big Bang model. Problems: Singularity; flatness; horizon; galaxy formation.
4. Inflation: Lack of a quantum theory of gravity. How "inflation" trys to solve the above cosmological problems. An emergent theory: The multiverses.

J. C. Mather: "The COBE project: Achievements in perspective"

To begin with I would like to point out that we can look back in time by looking at things that are very far away. We simply use the fact that light comes to us at a definite speed and so when we look at the sun, we see the sun as it was 500 seconds before we look at it; when we look at something in the middle of the galaxy we see it as it was 25 000 years ago and when we look at something extremely far away we can see it as it was 15 billion (10^9) years ago.

> *General considerations: No observable center / Hubble's law / Friedmann universes / Dark matter / Galactic rotation / Multiple images of distant galaxies...*
> *Main events: Nucleosynthesis / Number of photons fixed / Decoupling / Galaxy formation...*
> *The three COBE instruments: Differential Microwave Radiometer (DMR) / Far Infrared Absolute Spectrometer (FIRAS) / Differential Infrared Background Experiment (DIRBE)... COBE's scientific team...*

To conclude I would like to talk a little bit about what comes after COBE. We are going to turn off the COBE satellite this December because we got the data we wanted to get. We still have experiments

to look for the anisotropy at the most interesting scale currently, 1°
or 1/2° in size. People are doing balloon experiments going to the
Antarctica, making interferometers to use on the ground, and even
thinking about sending additional space probes out to get away from
the interference of the Earth's atmosphere.

J. Lejeune: "Neurophysiology of human intelligence"

Were the human intelligence the fruit of pure chance, how could it
be able to decipher, however partially, the rules of the universe?

A connivence between the physical/chemical laws and the pecu-
liarities of the substratum of the thought is worth investigating.

The inborn circuitry of our neurons could be our teacher in
elaborating theories. From Euclid's ideal plane to Cartesian coordinates, from Galileo's falling marble to Newton's attraction, from
Maxwell's little devil to Einstein's space-time, the milestones of science have their preexisting anatomical counterparts, macroscopic or
microscopic.

For example, the succession of the mathematical discoveries
occurred in the same sequence as the steps we use in analyzing optical information from the outside world down to the obscure center
that sees.

Discovery or reminiscence, the world of the ideas is not in dishar-
mony with biological observation.

The great caution remains: reason's networks are only one of the
tools at our disposal. The heart, with its built-in memory of life,
should never be overlooked. Separating them (heart and reason) is a
deadly blow to intelligence itself.

H. Elsaesser: "The Via Lactea"

The investigation of the *Via Lactea*, our galaxy, is a classic subject of
study in astronomy with a long history. However, new and interesting aspects come out again and again in such a way that several of
the problems of maximal actuality today, such as the active galactic
nuclei or the search for dark matter, are subjects of equally active
research on the *Via Lactea*, as older problems.

A summary in a few words of what astronomy has learned from research on our galaxy could be the following:

– What are the astral bodies (their physical nature, their formation, their development...)?
– What is a galaxy (its star content, its dimension, its structure...)?

Today there can be no doubt that the system which makes up the *Via Lactea* is one of the large disk-shaped spiral galaxies similar to Andromeda and to a multitude of other like systems which we observe in the universe.

> *Local Group / Generations of stars / Galactic structure / The Hertzsprung–Rusell diagram / Star age at the time of separation from main sequence / M17 and M13 / Distribution of young objects along the Via Lactea spiral / IR measurement on the Via Lactea with IRAS / Spiral structure / Distribution of giant gaseous regions / Orbit of Spherical groups M15 / Velocity rotation as a function of distance to galactic center...*

What type of matter constitutes dark matter? It is not clear. Maybe a multitude of white dwarfs and red stars with brightness too low to be detected by the methods presently used. Another hypothesis is neutrinos with mass or even unknown elementary particles.

M. Catalan: "The Planet Earth"

During the last few decades joint developments in radiostronomy, astrophysics and astronomy have given rise to a complete practical reconsideration of what we know about the universe.

Similarly, the models, on which our knowledge of the formation, interaction and physical structure of our planet were based, have undergone not only profound changes but a veritable revolution of the theories, and concepts that up to now are generally accepted have been substituted by opposite concepts and theories.

Then, for the first time, we can approach a physical picture of the evolutive dynamics of the Earth as a part of a universe in continuous evolution.

In this lecture, the formation of the Earth within the interior of the stellar system is described; the process of formation and dynamic evolution is examined; the special techniques available from the Earth and from outer space using today's astrometry can be used to observe the geometry of the universe through the deformation of extragalactic reference systems.

J. A. Gonzalo: "From the solar system to the confines of the observable universe"

Human ingenuity was able, in ancient Greece, of going from an accurate estimate of the size of the earth to rough estimates of sizes and distances to moon and sun. In medieval times, at the University of Paris, the concept of inertial motion and the realization of the possible rotation of the Earth prepared the way for Newton and for the idea of a cosmic gravitational force.

In the twentieth century, Einstein's Theory of Relativity, the general recession of galaxies, the cosmic relative abundance of light elements and the discovery of the cosmic background radiation paved the way for the big bang concept, earlier envisioned by Lemaitre. Today we can get fair estimates of the scales of masses, distances, and times in the universe, confirmed by observation.

We can conclude that the universe is admirably well done and that human intelligence (not without strenuous effort sometimes accompanied by good luck) is admirably well suited to investigate it. Good astrophysics does not lead to God directly, but it certainly is a useful pointer to the Creator.

S. L. Jaki: "Cosmology: An empirical science?"

Cosmology has become a branch of physical science as avidly cultivated as, for instance, are electromagnetics and atomic physics. This is in a marked difference from the status of cosmology a hundred, let alone two hundred years ago. Still a basic and hardly ever noted difference remains between scientific cosmology and other branches of physics. Unlike these branches, which can empirically demonstrate the existence of their subject matter, scientific cosmology has not yet done so. Its cultivators still have to ponder in depth some such questions: Is the universe the largest sum of galaxies that can actually be observed? Is the universe that sum of material entities whose existence can be inferred on a theoretical basis? Or is the universe an absolute totality of matter? Can such an entity be ever the object of a scientific cosmology which is unambiguously empirical? If not, what are the considerations, be they non-empirical and non-mathematical (and in that sense non-scientific), that can assure one about the existence of such a totality?

In 1993, at the time of the El Escorial conference, the age of the universe was located somewhere between ten and twenty billion years. And the numerical value of Hubble's parameter was estimated between 50 and 100 km/s/Mpc. The interaction with Ralph Alpher, John Mather, Stanley Jaki and others at the conference (and after) was extremely useful in order to make later accurate anticipations of $t_0 = 13.7 \cdot 10^9$ yrs and $H_0 = 65$ km/s/MPC. These accurate estimates were published in 1998 at *Acta Cosmologica (Krakow)*, with the help of my dear Polish friend Dr. Janusz Przeslawski, and some years later reprinted as an Appendix in *Inflationary Cosmology Revisited* (World Scientific, Singapore, 2005).

My colleagues N. Cereceda, G. Lifante and myself did not need any of the artificial assumptions of Inflationary Theory to arrive to those accurate numerical estimates, beautifully confirmed in 2003 by WMAP's data, the NASA satellite successor of COBE.

I would like to thank Manuel M. Carreira SJ Gines Lifante, and Manuel Marques for their encouragement and competent criticism. To Stanley Jaki and Ralph Alpher (both recently deceased) as well as to John C. Mather, Dermott Mullan and Anthony Hewish for their helpful comments and suggestions, and to my former student and collaborator Feli Martínez Ruiz for typing and reviewing the manuscript.

I am grateful also to Prof. K. K. Phua for his editorial advice and generous support.

Julio A. Gonzalo
Madrid, May 31, 2010

Foreword to the Second Edition

An APS editorial entitled "Gravitational waves detected by LIGO" was circulated on 12 February 2016. In it Emmanuele Berti (Department of Physics & Astronomy, The University of Mississippi, USA, and CENTRA, Universidad de Lisboa, Portugal) reported:

> "Gravitational waves emitted by the merger of two black holes have been detected, setting the course for a new era of observational astrophysics."

According to this report LIGO's detectors will allow in the future studies of the space-time curvature due to strong field behavior at cosmological distances. Observations indicative of cosmic space-time curvature $k < 0$ might be a distinct possibility. These consequences are discussed in the final chapters of the present work.

LIGO's interferometers are just maximally improved "L" shaped Michelson interferometers like those used by the American Nobel Laureate in Physics in the 80's to measure the velocity of light in perpendicular directions. The interferometers are supplemented with Fabry–Perot cavities at both sides of the beam splitter and use an extremely powerful laser source. In this way the distance travelled by the light in both arms is increased from 4 km to 1120 km, and the power of the laser light to 750 kW. This contributes to improve drastically the interference pattern of the two beams superimposed from the beam splitter to the detector. With these improvements the observer can verify that a gravitational wave has been detected.

These interferometers are classified as Dual Recycled Fabry–Perot Michelson Interferometers. They are carefully protected from seismic vibrations with "active damping" and "passive damping" devices and work under an extremely high vacuum. Measurements of small cosmic vibrations of the order of 10^{-19} cm are possible with LIGO's interferometers.

The *Phys. Rev. Lett.* abstract reporting the detection of gravitational waves is reproduced below:

Observation of Gravitational Waves from a Binary Black Hole Merger
B.P. Abbot *et al.* (LIGO Scientific Collaboration and Virgo Collaboration)
Phys. Rev. Lett. **116**, 061102. Published 11 February 2016.

Abstract

On Sept. 14, 2015 at 09:50 UTC the two detectors of the laser Interferometer Gravitational-Wave Observatory simultaneously observed a transient gravitational wave signal. The signal sweeps upwards in frequency from 35 to 250 Hz with a peak gravitational-wave strain of 1.0×10^{-21}. It matches the waveform predicted by general relativity for the inspiral and merger of a pair of black holes and the ringdown of the resulting single black hole. The signal was observed with a matched-filter signal-to-noise ratio of 24 and a false alarm rate estimated to be less than 1 event per 203,000 years, equivalent to a significance greater than 5.1σ. The source lies at a luminosity distance of 410^{+160}_{-180} Mpc corresponding to a redshift $z = 0.09^{+0.03}_{-0.04}$. In the source frame, the initial black hole mases are $36^{+5}_{-4}M_\odot$ and $29^{+4}_{-4}M_\odot$, and the final black hole mass is $62^{+4}_{-4}M_\odot$, with $3.0^{+0.5}_{-0.5}M_\odot c^2$ radiated in gravitational waves. All uncertainties define 90% credible intervals. These observations demonstrate the existence of binary stellar-mass black hole systems. This is the first direct detection of gravitational waves and the first observation of a binary black hole merger.

Foreword to the Third Edition

In the third edition of this, which is almost coincident with the launch of NASA's James Webb Space Telescope (JWST), reports recent work discussing the evidence in favor of an open, finite universe (Appendix C), and a more in depth evaluation of the Heisenberg–Lemaitre time (Appendix D) which takes into account that the cosmic expansion velocity at very early times is $\dot{R}(y_{HL}) \gg c$, much higher than light velocity in a vacuum.

Contents

List of Tables

List of Figures

Part I: Facts and Principles

As a firm starting point we are left with only a few principles, among them the universal law of energy

Max Planck
(1858–1947)

Antrittsrede zur Aufnahme in die Academie vom 28 Juni 1894, Physikalische Abhandlungen, 3:4, p. 3

Chapter 1

Energy Conservation

One of the most dramatic achievements of nineteenth century physics[1] was the enunciation of the principle of *energy conservation.* Clearly, this fundamental principle has very profound consequences in every branch of physical science.

Nineteenth century physicists realized that burning coal could set a steam machine in motion; and that this motion was not in any sense a creation out of nothing, but the result of a direct transformation of energy from one form into another. Therefore, mechanical work could be done only because a certain amount of energy was previously stored into the burning material, but only up to a certain amount, and no more.

Further, the work of a steam-engine itself could be translated into electrical energy which, then translates into heat or light, both of which, again, into chemical or mechanical energy. The total amount of energy could not be increased or diminished, in any way whatsoever, in any of these transformations. This is the basic meaning of the principle of energy conservation.

Haeckel,[2] a prominent German nineteenth century materialist, dared to combine the principles of matter conservation and energy conservation to set forth what he called the fundamental principle of the Monist Religion of the future.

He meant by it that this fundamental principle was in direct contradiction with the Christian views, which saw man and cosmos as created by an intelligent and all-powerful Creator. His was, and still is, a gratuitous assertion: What doctrine of the Church did forbid accepting the principle of matter and energy conservation, if the evidence in its favor should seem conclusive? Should Christianity be reproached for not teaching this principle in the "Credo"?

Of course, Christian Revelation was not set forth to teach man physics or chemistry, or geology, or biology as such, i.e., to teach truths of the natural order. As no less an authority than Saint Augustine said about sixteen centuries ago, Revelation tells us, "not how heavens go, but how to go to heaven".

The whole four thousand years old Judeo-Christian tradition, and especially the two thousand years old Catholic tradition, is witness of a staunch defense of reason as a God given gift to man, so that man may be able of acquiring truths of the natural order. But reason should not be confused with revelation, which certainly can go beyond reason, but never directly against it to the extent of contradicting it flatly. Reason can help revelation. In fact it is clearly needed to properly understand Revelation. And Revelation, if not abused, and properly respected, may help to get insights into the natural order, insights which otherwise might be missed.

According[3] to Monsignor Georges Lemaitre (1894–1966), the father of the Big Bang model:

> *Both — the believer and the non-believing scientist — will tray to decipher the complex palimpsest of nature in which traces of the long evolution of the world are confused and mixed up with one another. The believer may have the advantage of knowing that the enigma has a solution, that the written text underlying it is, after all, the work of an intelligent being, given that the problem proposed by nature has been posed to be solved, and that, no doubt, its difficulty is proportioned to the capacity of men now or in the future. This will not give him, perhaps, more resources in his research, but it will contribute to maintain him in that healthy optimism without which no sustained effort will last for very long.*

Is it too much to ask from contemporary cosmologists full respect for the principle of energy conservation?

Sir Arthur S. Eddington, mentor of Georges Lemaitre and himself one of the greatest contributors to early twentieth century cosmology said:

> *We are of course allowed to rearrange the matter of the universe... but in such rearrangement the experimenter cannot, and the theorist must not, violate the conservation of energy.*

The universe we see today, through ground-based powerful telescopes and through space telescopes (like NASA's Hubble telescope), is made up of a spherical distribution of galaxies (truly enormous but finite), of the order of 10^{12}, receding radially from one another at very large speeds within an expanding, homogeneous and quasi-isotropic spherical cloud of radiation, both centered at a certain point in space-time, a point which could be properly called the center of mass of this finite universe.

13.7 Gigayears ago,[4] more or less, all cosmic matter and all cosmic radiation were concentrated at this singular point in space-time, the center of mass of the universe. The Big Bang must have taken place precisely then and there. Matter and energy may reasonably be assumed to be conserved thereafter up to present times and beyond. As will be seen below, the uncertainty principle does not allow us to say anything before a time of the order of 10^{-103} sec after the Big Bang.

Einstein's original[5,6] cosmological equation for a galaxy of mass M_G receding from the center at maximum speed, $\dot{R}$, located therefore at maximum distance of R from it, can be written as

$$\frac{1}{2}M_G\dot{R}^2 = M_G\frac{4\pi}{3}G\rho R^2 - \frac{1}{2}M_Gkc^2 + \frac{1}{2}M_G\frac{\Lambda}{3}R^2c^2.$$

The term on the left-hand side is the kinetic energy and the three right-hand side terms are, respectively, the *gravitational potential energy*, the (relativistic) *space-time potential energy*, and the so-called *cosmological constant potential energy*, originally introduced by Einstein to counter gravitation and to produce (what he then through was) a "static" universe. Because his previous success in applying general relativity to solve problems, such as the bending of stellar light passing close to the Sun's surface, were *all* cases in which the space-time curvature was positive ($k > 0$), he was expecting $k > 0$

also in his cosmological equation, and he needed a cosmological constant term to counter gravity and produce a static universe. It was realized later (after the discovery of the general galactic recession) that the second term with $k < 0$ instead of $k > 0$ could play basically the same role as a third term with $\Lambda > 0$ in Eq. (1.1). Einstein, then, disavowed his cosmological constant. In his letter to Lemaitre[7] dated September 26, 1947, he said:

> *Since I have introduced the term (the "cosmological constant" term) I had always a bad conscience. But at the time I could see no other possibility to deal with the fact of the existence of a finite mean density of matter. I found it very ugly indeed that the field law of gravitation should be composed of two logically independent terms which are connected by addition ... I am unable to believe that such an ugly thing should be realized in nature.*

Lemaitre, on his part, did not agree with Einstein, and saw in this "cosmological constant" term perhaps a reasonable escape to a serious problem he then had: trying to explain a too-short cosmological age of the universe based upon the current Hubble's wrong estimate of the galactic distance scale. The problem was definitely solved half a century later when better determinations of the scale of galactic distance made compatible dynamic cosmological estimates of the age of the universe, and with the ages of the oldest globular star clusters known (turning off main sequence).

As it will be shown in greater detail in Part II, the *k-term* (space-time curvature term) and the Λ-*term* (cosmological constant) can be shown to play basically similar roles in the solutions of Einstein's cosmological equations. Therefore, it is adequate to proceed, as did Friedmann, setting $k < 0$ and $\Lambda = 0$, or equivalently, setting $k = 0, \Lambda > 0$, in order to obtain the dynamic evolution of an open universe. However, as also pointed out in Part II, $k < 0$ can be given a straightforward physical meaning linked to the radiation pressure associated to the cosmic background radiation, which is not so easy with $\Lambda = 0$.

It will be seen that dividing both sides of Eq. (1.1) by $\frac{1}{2}M_G\dot{R}^2$ we get

$$1 = [\Omega_m(t) + \Omega_r(t)] + \Omega_k(t) + \Omega_\Lambda(t). \tag{1.1}$$

Here $\Omega_m(t) = \rho_m(t)/\rho_c(t)$ is the dimensionless *matter* mass density parameter, obtained dividing the actual (time-dependent) matter mass density by the so-called critical (also time-dependent) matter mass density, corresponding to $k = 0$; $\Omega_r(t)$ is the radiation mass density parameter obtained likewise, which is at present much lower, but it was equal to it at decoupling (matter/radiation equality, atom formation time); $\Omega_k(t)$ is the time-dependent space-time *curvature* density parameter, and $\Omega_\Lambda(t)$ is the cosmological constant density parameter, which can be ignored for the moment for practical purposes, as did Friedmann. We can set $\Omega = \Omega_m + \Omega_r$ for convenience.

It will be seen that, using the compact Friedmann–Lemaitre solutions for an open $k < 0, \Lambda = 0$ universe, we get

$$t = \frac{R_+}{c|k|^{1/2}} [\sinh y \cdot \cosh y - y] \tag{1.2}$$

$$R = R_+ \sinh^2 y \tag{1.3}$$

with $2G(\frac{4\pi}{3}R^3\rho) = \text{constant} = c^2|k|R_+$. These solutions lead to the value of $\Omega(t)$ coming down from $\Omega(0) = 1$ (at very early times) to $\Omega(t_o) \approx 0.044$ (at present), and for $\Omega_k(t)$, coming up from $\Omega_k(0) = 0$ to $\Omega_k(t_o) \approx 0.965$ at present.

It will be seen below that this is relevant for finding a reasonable explanation for the large missing dark mass and the large missing dark energy problems.

Energy conservation from the very beginning to the present is of course compatible with a change in the equation of state of the universe at the transition from a *plasma universe* made up of nuclei (mainly H and ^{4}He) and electrons, to an *atomic universe* made up of atoms, then made up of cosmic dust, and then of stars and galaxies, as discussed in Part III.

References

1. K. A. Kneller, *Christianity and the Leaders of Modern Science*, with an introductory essay by Stanley L. Jaki (Real View Books, Fraser, Michigan, 1995), p. 7.
2. E. Haeckel, *Die Zukunft III.* (Berlin, 1895), p. 199.

3. O. Godart-M. Heller, *Les Relations Entre la Science et la Foi Chez Georges Lemaitre*, in Pontificia Academia Scientiarum Commentarii, Vol. III, No. 21, p. 11.

4. J. A. Gonzalo, *Inflationary Cosmology Revisited* (World Scientific, Singapore, 2005), Appendix, p. 83.

5. S. Weinberg, *Gravitation and Cosmology*, (Wiley and Sons, New York, 1972).

6. S. Weinberg, *Cosmology* (Oxford University Press, Oxford, 2008).

7. J. Farrell, *The Day Without Yesterday* (Thunder's Mouth Press, New York, 2005), p. 169.

Chapter 2

Energy Non-Conservation Means Too Much Freedom

We may begin by noting that the principle of energy conservation has not been consistently honored by some prominent physicist from very early in the twentieth century.

Both Planck and Einstein, the two greatest and most creative physicists of the past century,[1] were both realists and objectivists, as opposed to sensationists or purely subjectivists, as was Ernst Mach (1838–1916), who influenced both of them early in their scientific carriers and most of his followers. But Mach was soon left behind by Planck and Einstein when, forced both by their respective discoveries (the Theory of Quanta and the Theory of Relativity) they embraced realism as their philosophy of the physical world.

According to Planck,[1] speaking about the great pioneers: "What moved them (Copernicus, Kepler, Huygens, Newton and Faraday) was their firm belief in the reality of their picture (of the world) whether founded on an intellectual or on a religions basis." And, in another occasion, he wrote:

> As a firm starting point we are left with only a few principles, among them the universal law of energy.

9

Einstein[1] in a letter to his friend Maurice Solovine said:

> *You find it surprising that I think of the comprehensibility of the world (insofar as we are entitled to speak of such world) as a miracle or an eternal mystery. But, surely, a priori, one should expect the world to be chaotic, not to be grasped by thought in any way. One might (indeed should) expect that the world evidenced itself as lawful only so far as we grasp it in an orderly fashion. This would be a sort of order like the alphabetical order of words. On the other hand, the kind of order created, for example, by Newton's gravitational theory is of a very different character. Even if the axioms of the theory are posited by man, the success of such a procedure supposes in the objective world a degree of order which we are in no way entitled to expect a priori.*

In 1930 Wolfgang Pauli postulated the existence of the neutrino, a neutral particle of negligible mass, to justify the apparent loss of energy and linear momentum in the β disintegration process of neutrons:

$$n \to p + \bar{e} + \bar{\gamma}_e.$$

Being a neutral particle and having a negligible mass, if any, the neutrino would be extremely difficult to detect. Niels Bohr, who had early proposed the pioneering model for the hydrogen atom, and was then one of the most respected physicists in the world, did not believe in the neutrino's existence, and even went as far as saying that, perhaps, in the microscopic world of the atom, energy conservation was not an absolute requirement. This sounds a little strange in a man of his scientific caliber. But it was influential in discrediting for 25 years Pauli's proposal. In 1956, Clyde Cowan and Frederick Reines (who, many years later was awarded the Physics Nobel prize, which was also many years after the death of Cowan) demonstrated experimentally the existence of neutrinos bombarding pure water with a beam of 10^{18} neutrinos per second and subsequently detecting the photons emitted, in a very careful and laborious experiment: they had discovered the electron neutrino.

If needed, this was a most striking confirmation of energy conservation in the subatomic realm.

In 1987, Leo Max Lederman, Melvin Schwartz and Jack Steinberger discovered[2] the other two types of neutrinos, today known as the tau-neutrino and the muon-neutrino.

For some strange reason, in the physical realm of the whole cosmos, physicists have been careless with energy conservation. When, in 1948, Gold, Bondi and Hoyle[3] put forward their Steady State Model against the Big Bang Model, they had recourse to the *"continuous creation of matter"* in order to picture the general recession of galaxies in an unchanging *time-independent* universe. According to Hoyle[4] the idea of "continuous creation" can be described as "matter chasing its own tail", a fact in which he finds more "sweeping grandeur" than in the account of creation in Genesis, an account which is *not* meant, evidently to be a *scientific* description of anything, but which does not lack "grandeur" surely.

Apparently, more than a few distinguished theoretical physicists, including some Nobel Prize winners, did not seem to care very much about energy conservation when they did not appear to see anything wrong with the Steady State Theory.

Many years later, in 1980, Alan Guth[5] formulated his theory of cosmic inflation, which postulated an extraordinary (not conserving energy) cosmic expansion *at constant density*, taking place some 10^{-39} s after the Big Bang. As Guth says, the success of the Standard Model for elementary particles had produced a great impact within theoretical physicists. The fast pace of developments and the striking experimental confirmations by the late 70's resulted in the confident announcement that a full unification of all physics was at hand.

Grand Unification Theories (GUTs) predicted that the strong, weak and electromagnetic interactions would be unified at high energies ($E \approx 10^{16}$ GeV) and high temperatures ($T \approx E/k_B \approx 10^{29}$ K). These theories did not provide any hint as to why such high temperatures were needed to produce unification (symmetrization) of the interactions. Guth postulated a first-order (discontinuous) *phase transition* at still-higher temperatures ($T \approx 10^{31}$ K) at which the gravitational interaction would unify with the others. The phase transition would be analogous to the liquid–vapor transition, usually accompanied by supercooling, but quite different, in other respects, from phase transitions in the real world, which always conserve energy. According to Guth,[5] inflation (the ultimate in free lunch) takes place at constant density, between 10^{-37} s and 10^{-35} s, and results in an enormous growth in the cosmic radius, from $R = 10^{-52}$ m to $R \approx 2$ m.

In 1993, George Smoot gave an excellent talk at El Escorial on the recently measured CBR anisotropies detected by NASA's COBE satellite. He was talking at a Summer Course,[6] which I had organized together with Ignacio Cantarell and Rodolfo Nuñez de las Cuevas. At the end of George's talk, I made a comment: The equations in Inflationary Theory look very much like some equations in the old Steady State Theory resulting in growth at a constant density, but, now, conveniently back in time, some 10^{-37} s after the Big Bang, well beyond any possibility of being observed. His answer was that Inflation was not exactly the same as the Steady State Theory: a correct but perhaps insufficient answer.

Later, Vilenkin[7] summarized the story Alan Guth originally introduced his Inflationary Theory:

> Guth was telling us that the veil of mystery surrounding the big bang could be finally lifted. His new theory would uncover the nature of the big bang and explain why the initial fireball was so-contrived. The seminar room fell suddenly silent. Everybody was intrigued...
>
> Everything reduces to this: "tunneling" from "nothing" to "something". A half physical, half metaphysical argument, deviced (apparently seriously) to dispense with the requirement of a free Supernatural Agent, creator of the cosmos, i.e., of a "natural" system ruled by "natural" laws.
>
> Apparently Guth and other cosmologists were looking for "marketable" ideas. And they succeeded. Like Dali and Picaso had succeeded before them.
>
> The Big Bang theory was increasingly identified thereafter with the end of inflation. "Finetuning" became instantly popular. The cosmological constant was identified with "the full energy density of vacuum". After a number of clever considerations "we conclude that quantum processes during inflation necessarily generate a distribution of regions of all possible values of the cosmological constant."
>
> And if the fundamental particle physics allows the constants to vary, then quantum processes during eternal inflation inescapably

generate vast regions of space with all possible values of the constants. Eternal inflation thus provides a natural arena for applications of the anthropic principle.

In other words: too much "freedom".

It looks that the "freedom" denied to the Creator is somehow usurped by unscrupulous cosmologists in order to design literally countless imaginary worlds.

Quantum Mechanics is itself a "statistical" theory. Therefore it is not apt to describe a unique event, like the Big Bang. The Theory of Multiverses, so popular now for some time, is only a clever artificial concept, made up to jump from the real physical world we know to an imaginary multi-world existing only in our minds.

References

1. S. L. Jaki, *The Road of Science and the Ways to God.* (The University of Chicago Press, Chicago, 1978), Chapters 10 and 11, pp. 178, 192.
2. L. M. Lederman, *New Scientist*, 29 October 1988, p. 30.
3. H. Bondi, Spherical symmetrical models in general relativity, *Mon. Not. R. Astron. Soc.* **107** (1947) 410.
4. F. Hoyle, *The Nature of the Universe* (Harper Row, New York, 1950).
5. A. Guth, *The Inflationary Universe* (Perseus Books, Cambridge, Massachusetts, 1997), p. 187.
6. J. A. Gonzalo, J. L. Sanchez Gomez and M. A. Alario (eds.), *Cosmología Astrofísica* (Alianza Universidad, Madrid, 1995).
7. A. Vilenkin, *Many Worlds in One* (Hill & Wang, New York, 2006), pp. 10, 11, 137, 138.

Chapter 3

The Four Interactions

It is remarkable[1] that there are *four* and only *four* fundamental inter-actions in the universe. All stable physical entities from nuclei to galaxies are held together by a combination of the four interactions: the electromagnetic, the strong nuclear, the weak nuclear and the gravitational. Two decades ago, more or less, the *Physical Review Letters* published a number of papers claiming evidence for a *fifth* interaction, analogous to but different from the conventional gravitational interaction, which was finally discarded for lack of conclusive evidence.

In a solid for instance, *attractive* forces (of electromagnetic nature) must be present. Otherwise the solid would evaporate. And, at the same time, *repulsive* forces (also of electromagnetic nature) must be opposing the attractive forces to reach equilibrium. Otherwise the atoms making up the solid would collapse. The same type of equilib-rium must be achieved in a star, like the Sun, between the enormous gravitational attraction pulling the individual particles toward the center of mass and the huge repulsion, due to radiation pressure gen-erated by fusion at its core, pushing them away. In principle, one should say something similar about the elementary particles. If they have a finite size, as they do, the matter constituting the particle should be subject to competing attractive and repulsive interactions balancing each other, be they charged or neutral.

The *electromagnetic interaction* governs much of everyday life on our planet. All inorganic and organic chemical processes involve interactions between atoms, or ions, or molecular units which are of electric or magnetic character. Natural processes, like storms and hurricanes, are driven basically by electromagnetic interactions in the atmosphere, in the Earth's surface or in the oceans. Industrial processes, everyday life in large cities, transportation, communications, and agriculture, all depend to a large scale also on physical and chemical processes which are also driven by electromagnetic interactions.

The *strong nuclear* interaction is responsible for holding protons and neutrons together in atomic nuclei. It is strong enough to hold together many positively charged protons within heavy nuclei which repel each other strongly. The strong nuclear interaction acting among nucleons (protons and neutrons) is strong enough to bind uranium nuclei, with 92 protons, by adding 143 neutrons.

In spite of its short range, the strong interaction is therefore of the order of one hundred times stronger than the repulsive electromagnetic interaction. If it were one half weaker[2] the Periodic Table would contain substantially less stable elements, and life on Earth would not be possible. If it were twice as strong, the lifetime of midweight stars, like the Sun, would be much shorter.

The *weak nuclear* interaction, roughly 10^{-12} times weaker than the strong nuclear interaction, is important in many respects. For instance, it governs the spontaneous conversion of free neutrons in protons, and the inverse process. When a massive star explodes as a supernova,[2] the weak interaction is responsible for the energy deposited by neutrinos on the expanding shock front. The weak interaction precipitates the initial collapse, which allows the return of the heavier elements in the outer layer of the collapsing star to the interior of the galaxy.

In this way, for second or third generation stars like our Sun, their planets can have enough heavy elements essential for life. The weak interaction is also crucial in producing ^{4}He soon after the Big Bang, which plays an important role in the formation of stars.

The gravitational interaction, attractive always, by far the weakest at small scales, becomes dominant at large scales. The almost-negligible gravitational interaction, in large bodies (planets, stars, galaxies *etc.*), overwhelms completely the other interactions. Stars are formed out of galactic hot gas when gravity overcomes the gas pressure forces and the turbulence causing it to coalesce and contract. With a gravitational interaction 10^6 times stronger, according to Martin Rees[3]:

> *The number of atoms needed to make a star (a gravitationally bound fusion reactor) would be a billion times less... in this hypothetical strong-gravity world stellar lifetimes would be a million times shorter. Instead of living for ten billion years, a mini Sun would burn faster, and would have exhausted its energy before even the first steps in organic evolution had got under way.*

Such a star would have about one thousandth the luminosity, three times the surface temperature and one twentieth the density of the Sun, burning too hot and too quickly. The opposite would be true for a universe with a gravitational interaction substantially weaker.

Since the beginning of the twentieth century physicists begin to characterize the fundamental interactions in terms of *universal constants*: $\hbar$ (Planck), k_B (Boltzman), c (velocity of light), G (Newton), e (electron charge), m_e (electron mass), m_p (proton mass), m_n (neutron mass, slightly larger than proton mass), τ_n (free neutron half-life) and others, often combinations of the ones mentioned.

Planck noted that it is possible to define "natural" physical units by means of combinations of these universal constants. For instance

$$m_{\text{Pl}} = (\hbar c/G)^{1/2} = 2.17 \times 10^{-5}\,\text{g}, \qquad (3.1)$$

$$l_{\text{Pl}} = (\hbar G/c^3)^{1/2} = 1.61 \times 10^{-33}\,\text{cm}, \qquad (3.2)$$

$$t_{\text{Pl}} = (\hbar G/c^5)^{1/2} = 5.36 \times 10^{-44}\,\text{s}. \qquad (3.3)$$

Table 3.1.   Cosmic quantities in Planck's units.

Cosmic quantity	Natural unit	Ratio
$M_u = 1.547 \times 10^{54}$ g	$m_{\text{Pl}} = 2.17 \times 10^{-5}$	7.13×10^{58}
$R_+ = 4.588 \times 10^{26}$ cm	$l_{\text{Pl}} = 4.588 \times 10^{-33}$	2.85×10^{59}
$t_+ = 1.152 \times 10^{16}$ s	$t_{\text{Pl}} = 5.36 \times 10^{-44}$	2.15×10^{59}

The *relative intensities* of the fundamental interactions[4] can be given in terms of the following dimensionless quantities:

Gravitational interaction:

$$\alpha_G = \frac{Gm_p^2}{\hbar c} = 5.9 \times 10^{-39}$$

Electromagnetic interaction:

$$\alpha_E = \frac{e^2}{\hbar c} \approx \frac{1}{137}$$

Strong nuclear interaction[5]:

$$\alpha_{\text{NS}} = \frac{g_s^2 e^{-r/r'}}{\hbar c} \approx 1 \quad (g_S^2/\hbar c \approx 15; r' \approx 1.5 \times 10^{-13} \text{ cm})$$

Weak nuclear interaction[5]:

$$\alpha_{\text{NW}} = \frac{g_W^2 e^{-r/r''}}{\hbar c} \approx 10^{-12} \quad (g_w^2/\hbar c \approx 15 \times 10^{-12};$$

$$r'' \approx 1.5 \times 10^{-15} \text{ cm})$$

The dimensionless quantities introduced above are given for convenience in terms of the proton mass m_p and the charge e, but they could be given in terms of the Planck monopole's mass. For instance

$$\alpha_G(m_{\text{Pl}}) = \frac{Gm_{\text{Pl}}^2}{\hbar c} \approx 1, \tag{3.4}$$

where m_{Pl} is the massive hypothetical particle in which the gravitational interaction and the electromagnetic interaction are of the same order.

It is interesting to note that at the time stars and galaxies begun to form in the universe, at a cosmic radius $R_+ = 2GM_u/c^2$, the main cosmic quantities can be given in terms of Planck's natural units and the numbers for M_u (cosmic mass), R_+ (cosmic radius) and t_+ (cosmic time) come close to each other.

It may be noted that M_u (cosmic mass)$/m_p$ (proton mass) gives the total number of baryons in the universe: 9.26×10^{80}.

Several physicists, including Dirac, Dicke, *etc.*, have considered the possibility of varying the universal constants with time through cosmic history. Up to now, the results have not been very conclusive.[6,7] In particular, the time-variation of Newton's gravitational constant, estimated from ancient eclipse records, and by various other means, is

$$(\dot{G}/G)_o \leq \text{between } 10^{-10} \text{ and } 10^{-11} \text{ per year};$$

the time-variation of α, the electromagnetic fine-structure constant, estimated from measurements on uranium ores in the Oklo river (Gabon) is

$$(\dot{\alpha}/\alpha)_o \leq 10^{-6} \text{ per year},$$

which implies either that e^2, $\hbar$ and c remain constant, or that they change simultaneously precisely to keep α almost constant, which does not seem very realistic.

References

1. R. Jastrow, *Red Giants and White Dwarfs* (Warner Books, New York, 1980), p. 46.
2. G. Gonzalez and J. W. Richards, *The Privileged Planet* (Reguery Publishing, Washington, 2004), pp. 201, 202.
3. M. Rees, *Just Six Numbers: The Deep Forces that Shape the Universe* (Basic Books, New York, 2000).
4. J. A. Gonzalo, *The Intelligible Universe* (World Scientific, Singapore, 2008).
5. R. Eisberg and R. Resnick, *Quantum Physics of Atoms, Molecules, Solids, Nuclei and Particles* (John Wiley: New York, 1947), pp. 689, 700.
6. F. Hoyle, *Frontiers of Astronomy* (London, Heinemann, 1955), p. 304.
7. H. Fritzsch, *The Fundamental Constant* (World Scientific, Singapore, 2009), p. 149.

Chapter 4

Matter, Radiation, Particles

Up to the beginning of the twentieth century, matter and radiation were considered as distinct entities. Matter, as making up a solid, a liquid, a gas or a plasma (a fluid of ionized atoms and electrons), was seen as altogether different from electromagnetic radiation, which spans from radio waves to X-rays and beyond. It may be noted that in a hot dense plasma, like the one making up a star, as our Sun, the electrons, which are almost two thousand times less massive than the protons or the ^{4}He nuclei, occupy much more volume than the much heavier protons and alpha particles. To begin with, the Compton wavelength of the electron,

$$\lambda_c(\text{electron}) = \frac{h}{m_e c^2} \approx 2.41 \times 10^{-10} \, \text{cm}, \qquad (4.1)$$

is much larger than the Compton wavelength of the proton,

$$\lambda_c(\text{proton}) = \frac{h}{m_p c^2} \approx 1.31 \times 10^{-13} \, \text{cm}, \qquad (4.2)$$

but one must take also into consideration Pauli's exclusion principle, which makes the effective electron radius in the plasma more like Bohr's radius[1]

$$a_0(\text{electron}) = \frac{\hbar^2}{m_e e^2} \approx 5.29 \times 10^{-9} \, \text{cm}, \qquad (4.3)$$

almost twenty times larger than the electron Compton wavelength. This is true also for the plasma in a star and must be true too for the universe in its early plasma state, just after the primordial electrons could be formed, and well before decoupling (atom formation).

Arthur Holly Compton (1892–1962) discovered in 1923 that bombarding paraffin with X-rays they are scattered with a wavelength larger than the wavelength of the primary X-rays. This is called the Compton effect and, for its discovery, Compton was awarded, together with C. T. Wilson, the Physics Nobel Prize for 1927. The effect is not explicable with a purely undulatory conception of the electromagnetic radiation (X-rays or other), but it was explained, by Compton and by Peter Debye (later Chemistry Nobel Prize, in 1936), as due to the elastic collision between the radiation quanta (X-ray packages) and the electrons of the paraffin atoms. In these collisions energy and momentum were transferred from the radiation quanta (particles) to the electrons, respecting the principles of energy and momentum conservation.

Compton's views on the particle nature of radiation[1] were shared from the beginning by Maurice de Broglie, a French experimental physicist. His experiments and the analysis he made of them impressed very much his brother Louis, a student of history at the time, to such extent that Louis changed his career from history to physics. In 1924, Louis de Broglie presented his doctoral thesis to the Faculty of Science of the University of Paris arguing, originally and convincingly, in favor of the existence of matter waves. In essence, he was putting together Planck's ($E = \hbar\nu$) quantum theory and Einstein's relativity ($E = mc^2$). The originality of this thesis was immediately recognized, but, because of the lack of experimental evidence supporting it, was put aside for the moment by his physicist colleagues.

However, Albert Einstein recognized immediately its validity, and called the attention of other physicists to Louis de Broglie's work. In 1929 de Broglie was awarded for this discovery the Physics Nobel Prize.

Subatomic particles (not necessarily elementary) are characterized by a rest mass (m), a radius (r), a lifetime (τ), a charge (q),

Table 4.1. Subatomic particles.

	m(g)	r(cm)	τ(s)	q(esu)	T(K)
Neutron	1.67×10^{-24}	1.2×10^{-13}	887	0	3.88×10^{12}
Proton	1.67×10^{-24}	1.2×10^{-13}	∞ (?)	4.8×10^{-10}	3.88×10^{12}
Electron	9.10×10^{-28}	(5.29×10^{-9})	∞ (?)	4.8×10^{-10}	2.11×10^{9}
Planck's monopole	2.17×10^{-28}	1.66×10^{-33}	—	—	5.04×10^{31}

a spin (intrinsic angular momentum) and a characteristic temperature $(T = mc^2/2.8k_B)$ above which they cannot exist as stable particles.

Experiments at high energy accelerators in the 1970's demonstrated unambiguously that neutrons and protons were made up by even more elementary constituents — the quarks. These elementary constituents, as far as we know, are not made up of smaller particles and have not internal structure.

According to the standard model there are three generations of *elementary fermions* (spin 1/2) including six quarks, three neutrinos, three leptons, their corresponding antiparticles, and four *elementary bosons* (spin 1) which mediate the electromagnetic (γ), the nuclear strong (g) and the nuclear weak (Z° and W$^{\pm}$) forces. The neutron, for instance, is made up of three quarks: d(down) d(down) u(up), and the proton of another set of three quarks: u(up) u(up) d(down). But the quarks themselves are not stable in isolation outside the nucleons of which they are part of. They are structureless and have a definite rest mass and a fractional $\pm$ charge. To have a definite physical existence, however fleeting, outside the nucleon, they must have some definite size, and some definite radius, which is difficult to measure and sometimes estimated as 10^{-18}cm, much smaller than the neutron or the proton radius.

What are the *cosmological implications*? A *material particle* with a given rest mass can be considered as a "pack" of energy tied up around itself, occupying a certain volume in space-time. One is therefore entitled to assign to it an average or characteristic radius and a characteristic density.

For instance, for a proton, its characteristic radius is

$$r_p(\text{proton}) \approx 1.2 \times 10^{-13}\,\text{cm} \tag{4.4}$$

and its characteristic density

$$\rho(\text{proton}) \approx \frac{m_p}{\frac{4\pi}{3}r_p^3} \approx 2.3 \times 10^{14}\,\text{g/cm}^3. \tag{4.5}$$

In the universe, at present, in our local neighborhood, we have galaxies, stars, planets and comic dust, made up of *atoms*, including nuclei and electrons, and a relatively small amount of radiation, the cosmic background radiation (CBR). Going back in time, the universe was denser and hotter, but the physical conditions were still apt to have atoms, which, before the formation of stars and galaxies, were in the form of dense and hot cosmic dust, surrounded by a hot cloud of radiation. Going further back in time, when the radiation temperature was around 3000–4000 degrees Kelvin, the atoms were ionized and the universe was a hot plasma of protons, ^{4}He nuclei and electrons, not yet too dense. At still earlier times the plasma was really hot: those were the times at which primordial nucleosynthesis took place. Cosmic density was then so high that individual electrons could not be separate entities. Before then, the universe contained protons (not yet ^{4}He nuclei) and some neutrons which had not yet disintegrated spontaneously. Still further back in time, at a temperature of about 3.88×10^{12} K, the density of the universe was nuclear density. Before that time cosmic density was larger than the density of ordinary neutrons or protons, so neither neutrons nor protons could exist as separate entities. Consequently, it is not very realistic to talk about a soup of nuclear particles and antiparticles at those times, because then they could not exist as such. Gamow, Alpher and Herman, spoke about the "ylem" (primordial matter) before the quarks were known to exist. They considered a moment in cosmic history at which the universe was an enormously dense aggregate of neutrons (so the net electric charge was zero) permeated by an enormous amount of radiation. This was their starting point to investigate the origin of nuclei heavier than hydrogen. Today we know that, leaving aside the nuclei of ^{4}He, and traces of other light elements, all other nuclei are cooked up in successive generations of stars. But this is another issue.

At earlier cosmic times that are hotter ($T > T = 3.88 \times 10^{12}\,\mathrm{K}$) it is highly speculative to reconstruct the thermal history of the universe. But one thing can be said. Before reaching the *singularity* ($t = 0$, $R = 0$), Heisenberg's uncertainty principle for a universe with a finite mass,[2] say $M_u \approx 1.547 \times 10^{54}\,\mathrm{g}$, sets a limit to the time before which the speculation becomes meaningless.

We could call it Heisenberg's time

$$t_H \cong \frac{\hbar}{M_u c^2} \approx 7.54 \times 10^{-103}\,\mathrm{s}. \tag{4.6}$$

It may be noted that at these early times the cosmic radius $R(t)$ is given by Einstein's cosmological equation as

$$R(t) \approx \text{constant} \times t^{2/3}. \tag{4.7}$$

Then the growth of cosmic radius between $t = t_H$ (Heisenberg's time) and $t = t_{\mathrm{Pl}}$ (Planck's time) is given by

$$R(t_{\mathrm{Pl}})/R(t_H) = (t_{\mathrm{Pl}}/t_H)^{2/3} = 1.71 \times 10^{39}, \tag{4.8}$$

a number of the order of the inflationary factor used by Alan Guth ($\sim 10^{40}$), without any need of inflation.

References

1.   R. Eisberg and R. Resnick, *Quantum Physics of Atoms, Molecules, Solids, Nuclei and Particles* (John Wiley, New York, 1974), pp. 63, 110.
2.   J. A. Gonzalo, *Inflationary Cosmology Revisited* (World Scientific, Singapore, 2005), p. 55.

Chapter 5

The Dark Night Sky and Olbers' Paradox

Olber's paradox[1,2] is the riddle of the darkness of the night sky in conflict with the presumed infinity of the universe. If there were infinite stars, in infinite galaxies, in the universe, looking in any direction from the Earth's surface the line of sight should end at the surface of a star and then the whole night sky should look bright.

Since the German astronomer William Olbers (1758–1840) reformulated the dark night paradox in 1823 he and many after him have tried to save the presumption that the universe is infinite. Some did so for preconceived philosophical reasons, although not necessarily regarding Olbers. As Stanley L. Jaki points out[1] correctly in his carefully researched book, an infinite universe could readily pass for the ultimate entity, and serve thereby as a substitute for God.

According to R. H. Dicke (Princeton University): *"Professor Jaki has considered with great care the origins, history, and significance of this question and his scholarly, but interesting and readable book will be* the *definitive historical statement for years to come"*.

Jaki's book reprints, in the Appendices, two short essays by E. Halley,[3] one longer essay by J. P. Loÿs de Cheseaux[4] and Olbers' essay,[5] *On the Transparency of Space*. He also shows that Olbers' kept referring repeatedly to Halley's wrestling with the question, to such extent that the paradox could well have been named Halley's paradox.

The history of the ups and downs of Olbers' paradox through the nineteenth and twentieth centuries[1] is certainly illustrative of the fact that the awareness of the scientific community to key developments in physics does not grow with time properly and steadily at all. The "rediscovery" of Olbers' paradox in the mid twentieth century can be attributed to Bondi and Gold[6] (and somewhat later to Hoyle[7]) in connection with their Steady State Theory. Ten years later, in his Joule Memorial Lecture, H. Bondi makes the following claim:

> *In cosmology, which is the name given to the subject that investigates the large scale structure of the universe, we can put a precise date to the moment when it became a scientific subject and left the road of philosophical speculation. This date is 1826. In 1826, the German astronomer Olbers published a little investigation which, although I doubt whether he realized it, made cosmology a science.*

This inaccurate claim is no doubt exaggerated, but the importance of the question brought up by the recurrent paradox is certainly momentous.

Olbers, like most of his contemporaries, was convinced that the universe was infinite in extent and he gave some reasons, though not very satisfactory,[1] to show that the paradox of the darkness of the night sky could be somehow solved.

But the key to the solution of Olber's paradox would come with the Einstein General Theory of Relativity, through the assertion that the *total mass of the universe can only be finite*[8] if scientific considerations about the universe as a whole are to remain meaningful:

> *The total mass M of the universe, according to our view, is finite, and is in fact*

$$M = \rho \cdot 2\pi^2 R^3 = 4\pi^2 \frac{R}{k} = \pi^2 \left(\frac{32}{k^3 \rho} \right)^{1/2}. \tag{5.1}$$

> *Thus the theoretical view of the actual universe, if it is in correspondence with our reasoning, is the following. The curvature of*

space is variable in time and place, according to the distribution of matter, but we may roughly approximate to it by means of a spherical space.

The total mass M of the universe could have been given perhaps more properly for a spherical cosmos by means of

$$M = \rho \cdot \frac{4\pi}{3} R^3, \tag{5.2}$$

with ρ as the average density, including mass matter and radiation matter, and R as the growing and now very large, but finite, radius of the universe. But the important point is, as Einstein points out boldly, that M should be *finite* to avoid physical inconsistencies.

Today we know that there are 10^{11} to 10^{12} galaxies, each containing 10^{11} to 10^{12} stars with an average mass of the other of the Sun's mass, so that $M_u \cong 1.54 \times 10^{54}$ g. The best telescopes in the world, at Hawaii, Chile, Tenerife *etc.*, could see galaxies substantially fainter than the faintest proto-galaxies or quasars *actually* observed by ground-based telescopes — but they do not *see* anything beyond a certain distance. The same is true of the space telescope Hubble. There are strong indications (through IR telescopes) that galaxies further away from the Milky Way are spatially more closely packed than galaxies in our immediate neighborhood. And the finiteness of the speed of light puts a limit to the recession velocity, and therefore to the maximum distance of observable protogalaxies emitting the light which is just arriving at us now, about thirteen billion years after they being emitted. As will be seen later the recession velocity of these young galaxies relative to us is beginning to saturate (it cannot exceed the speed of light relative to us) and one might interpret this as an effective acceleration of the galaxies in our neighborhood with respect to the most distant (youngest) galaxies.

It is possible to make a rough estimate of the ratio of radiation flux due to the CBR, which is at $T \approx 2.72$ K and is filling the cosmic plasma sphere as seen from our galaxy now, to that due to the visible light *if the whole sky* were fully occupied by stars in every direction.

The ratio of a nearby star's surface area at a distance

$$d_{\text{star}} \approx 5 \text{ light years} = 4.7 \times 10^{18} \text{ cm}$$

to the Sun's surface area at a distance

$$d_{\text{sun}} \approx 8 \text{ light minutes} = 1.44 \times 10^{13} \text{cm}$$

is the order of

$$\frac{A_{\text{star}}}{A_{\text{sun}}} = \left(\frac{d_{\text{star}}}{d_{\text{sun}}}\right)^2 \approx 9.56 \times 10^{-12}. \tag{5.3}$$

The Sun's surface temperature ($T_{\text{sun}} \approx 6000\,\text{K}$) is roughly equal to the surface temperature of a typical star, but the emitting surface, as seen from the Earth, is of course much larger. Then the estimate of the flux ratio would be

$$\frac{\Phi_{\text{CBR}}}{\Phi_{\text{star}}} = \frac{\Phi_{\text{CBR}}}{\Phi_{\text{sun}}}\left(\frac{A_{\text{sun}}}{A_{\text{star}}}\right) = \left(\frac{\sigma T_{\text{CBR}}^4}{\sigma T_{\text{sun}}^4}\right)\left(\frac{A_{\text{sun}}}{A_{\text{star}}}\right) \approx 4.41 \times 10^{-3}, \tag{5.4}$$

using Stefan–Boltzmann's law.

In a similar way, the ratio of the CBR flux to the flux of visible light coming from a night sky *if* filled with galaxies (distance between two typical nearby galaxies: $d_{\text{gal}} \approx 2 \times 10^6$ light years $= 1.89 \times 10^{24}$ cm), each containing about 10^{11} stars like our Sun, would be

$$\frac{\Phi_{\text{CBR}}}{\Phi_{\text{gal}}} = \frac{\Phi_{\text{CBR}}}{10^{11}\Phi_{\text{sun}}}\left(\frac{d_{\text{gal}}}{d_{\text{sun}}}\right)$$

$$= 10^{-11}\left(\frac{\sigma T_{\text{CBR}}^4}{\sigma T_{\text{sun}}^4}\right)\left(\frac{d_{\text{gal}}}{d_{\text{sun}}}\right)^2 \approx 2.09 \times 10^{-3}. \tag{5.5}$$

Of course, stars and galaxies do emit in the visible, but the CBR — a microwave radiation — is *invisible*, so that the night sky appears completely dark in spite of being pervaded by the very faint CBR radiation.

An observer looking to a night sky full of stars would see a star's surface in every direction of sight and would be seeing therefore a bright sky. The fact that we see the night sky as dark means therefore that the number of stars, and correspondingly the number of galaxies, is *finite*.

According to the Big Bang picture, just at t_H (Heisenberg time), R_H (Heisenberg radius), the universe was a tremendously concentrated *dot* of space-time filled with primordial *matter* and *radiation*.

Outside, as far as we can say, there was nothing: no matter, no radiation, no space, no time. Rather than undergoing an "explosion", the universe was undergoing an "implosion". The cosmic dot did grow *inside* the initial cosmic sphere. After a few microseconds, the first baryons (neutrons) did begin to exist; a few minutes later, the first electrons and protons appeared, massively, and neutrons and protons begun to fuse quickly into ^{4}He nuclei and other light nuclei in very small amounts; then nucleosynthesis stopped in about thirty minutes; much later, about a few million years later, atoms formed, the universe became transparent matter and radiation *decoupled* and the CBR began to flow from the outer space-time spherical shell towards the center. The Milky Way was moving then and now at a relatively small fraction of the speed of light with respect to the center of the CBR sphere. Therefore it can be considered to be placed relative near that center.

Much later, at about four hundred million years, the first stars and protogalaxies began to form. At that time the characteristic temperature the CBR was already as low as $T_+ \approx 60\,\mathrm{K}$, the cooling down and the expansion continued, and, finally, the Sun (a second or third generation star) formed, and around it the solar system, within which our planet Earth, about 4.8 billion years ago (i.e., about 8.9 billion years after the Big Bang).

At present, galaxies are some million years far away from each other, the CBR has cooled to $T \approx 2.72\,\mathrm{K}$, and the night sky looks dark. In the distant future, galaxies will be much farther apart, the CBR will be much cooler than it is now, and the night sky will look even darker.

As Eddington[9] said:

Infinity is a mischief and as such should have no place in physics.

Discussions of Olber's paradox in terms of the Steady State Theory, which assumed an infinite and eternal cosmos, received a death-blow in 1992, when the results of COBE (Cosmic Background Explorer) were made public.

In the early 80's Waldorn[10] (quoted by Jaki in *The Paradox of Olber's Paradox*) assumed, as apparently Halley had many years before, in order to solve the paradox, considered that as ever larger spherical shells are considered, the density of matter should decrease. This is exactly the opposite of what cosmic data in the IR support[11] and opposite of what Einstein's cosmological equations predict about cosmic evolution from early times to the present.

The *eternality* and *infinity* of matter is a fundamental Marxist dogma and as such it must have had, and still has considerable influence not only in the Communist world but also outside it.

J. Gribbin, in connection with Olber's paradox, makes the following comment:

> *The most fundamental observation in all of science is that night follows day. This simple fact is enough to show that the Universe has not always existed, everywhere, in the form that we see it today. There must be an "edge" to the Universe as we know it.*

The only way of removing Olber's paradox is to view it in terms of *cosmic finitude*. Attempts to do otherwise will end in inconsistent reasoning. Evidently, Olber's paradox cannot become a fundamental fact for cosmology when the universe becomes just our universe: one of many multiverses, which will be unobservable forever.

As Jaki[1] says:

> *It is not mathematics, but philosophy (and, one would add, common sense "physical realism", JAG) that alone conveys the impossibility of an actually realized infinite quantity and provides thereby the real foundation for a scientific respectable treatment of Olber's paradox.*

References

1.   S. L. Jaki, *The Paradox of Olber's Paradox*, new edition (Real View Books, Pinckney, MI, 2000).
2.   E. H. Harrison, *Darkness at Night. The Riddle of the Universe* (Harvard University Press, Harvard, Massachusetts, 1987).
3.   E. Halley, *Philos. Trans.* **31** (1720) 22–26.
4.   J. P. L. de Cheseaux, *Traité de la Comete* (Chez Marc-Michel Bousquet & Compagnie, Lausanne and Geneva, 1744).
5.   W. Olbers, On the transparency of space, *Edinburgh New Philosophical Journal*, April–October 1826, pp. 141–150.
6.   H. Bondi and T. Gold, *Mon. Not. Roy. Astron. Soc.* **108** (1948) 252–270.

7. F. Hoyle, *Frontiers of Astronomy* (Heinemann, London, 1955), p. 304.

8. A. Einstein, in *The Principle of Relativity: A Collection of Original Memoirs on the Special and General Theory of Relativity*, eds. H. A. Lorentz, A. Einstein, H. Perrett and G. B. Jeffrey (Dover, New York), pp. 177–188.

9. A. S. Eddington, *New Pathways in Science* (The University of Michigan Press, Ann Arbor, 1959), p. 217.

10. R. A. Waldorn, Is the Universe really expanding? *Speculations in Science and Technology* **4** (Dec. 1981) 539–543.

11. Private communication or, off hand remark to the author by R. A. Alpher, around 1993.

Part II: Relativistic Cosmology

*It is a magnificent feeling to recognize the unity
of a complex of phenomena which appear to be things
quite apart from the direct visible truth*

Albert Einstein
(1879–1955)

Chapter 6

General Relativity and Cosmology

The field equations of the General Theory of Relativity[1,2] are in principle the right *dynamical equations* to describe cosmic evolution. The field equations are given in terms of the Ricci tensor $R_{\mu\nu}$ (which is the contracted Riemann curvature tensor), the cosmological constant Λ (originally introduced by Einstein and finally discarded), and the stress-energy tensor $T_{\mu\nu}$ for all forms of matter and energy excluding gravity. $R_{\mu\nu}$ and R are determined by the *metric tensor $g_{\mu\nu}$* together with its first and second-order partial derivatives. The metric tensor $g_{\mu\nu}$ is defined implicitly by

$$(ds)^2 = g_{\mu\nu}dx^\mu dx^\nu \tag{6.1}$$

under the assumption that space-time is locally Minkowskian (i.e., possessing the geometric properties of special relativity) and symmetric (i.e., invariant under the interchange of μ and ν), where x^μ and x^ν are four arbitrary coordinates describing fully the *curved space-time*.

The field equations are therefore a set of second-order nonlinear partial differential equations for $g_{\mu\nu}$. Initial and/or boundary conditions must be assumed in order to specify a unique solution.

Since the field equations are symmetric, there are in general ten independent equations, one for each independent $g_{\mu\nu}$. But, if the curved space-time is homogeneous and isotropic as it is assumed to be the case for the physical Universe, the number of independent equations reduces to two, and finally to one equation,

$$\dot{R}^2 = \frac{8\pi}{3} G\rho R^2 - kc^2 + \frac{\Lambda}{3} c^2 R^2 \tag{6.2}$$

where R is the expanding cosmic radius, $\dot{R}$ its time derivative, G Newton's gravitational constant, ρ the mass density (of matter and radiation), k the space-time curvature, which can in principle be $k > 0$ (reinforcing gravity), $k = 0$ (neither reinforcing nor countering gravity), or $k < 0$ (countering gravity), c the speed of light in vacuum, and Λ the cosmological constant. Note that k is dimensionless, and Λ has the dimension of an inverse-square length. As mentioned, Λ was originally introduced by Einstein to counter gravity in order to get a static universe.

This probably means that the author of general relativity was thinking then in a $k > 0$, reinforcing gravity. When Friedmann[3] found the correct solutions for a *spherical homogeneous* and *isotropic* universe, he assumed $\Lambda = 0$, and saw that $k < 0$ had in fact an effect opposite to gravity, resulting in an expanding universe.

The original theory of gravitation, as it is well known, was formulated by Newton implying an instantaneous interaction between massive bodies. This is equivalent to assuming that the gravitational interaction propagates at infinite speed. To correct this shortcoming, Einstein formulated in 1916 his General Theory of Relativity, treating the gravitational interaction in a fully consistent way. The *principle of equivalence* says that an observer cannot distinguish between being in a system of reference which is subject to a gravitational field and being in one which is uniformly accelerated. This is tantamount to say that *inertial* mass and *gravitational* mass are equal. Physical laws must be such, according to Einstein, that they apply to any reference system regardless of its motion with respect to the mass distribution in the universe. Einstein took recourse to *curved spaces*, as introduced first by Gauss, Bolyai and Lobachevski (for the case of a *two-dimensional* space of constant curvature) and later generalized by Riemann (for *more dimensions*), to develop the complex mathematical formalism that is appropriate to describe relativistically the dynamics of accelerated systems.

But (and this is very important) *if* we take into account that the actual geometry of the universe is *spherical,* and that the cosmic distribution of matter (galaxies) and radiation (the CBR) is basically *homogeneous* and *isotropic,* the geometry of cosmic space-time reduces to that of a set of two-dimensional surfaces describable by a single curvature parameter K, which can be

$$K = \begin{cases} \dfrac{1}{a^2} \text{ (closed spherical surfaces with } positive \text{ curvature)} \\ 0 \text{ (plane, Euclidean space)} \\ -\dfrac{1}{a^2} \text{ (open Gauss, Bolyai, Lobachevski surfaces with} \\ \quad negative \text{ curvature),} \end{cases}$$

where, if a (the scale factor) satisfies $a \gg 1$, the closed and open symmetries became indistinguishable in practice.

Note that Einstein's cosmological equation Eq. (6.2) involves only one space-time curvature k, plus the artificial, and in fact unnecessary, counter-gravitational term Λ. Note also that the integration of Eq. (6.2) becomes insensitive to the values of the curvature $|k|$, for

$$R \ll R_+ = 2GM/|k|c^2$$

Previous to his application of the General Theory of Relativity to describe cosmic dynamics, Einstein had applied it successfully to predict such small gravitational effects as the bending of light rays in the Sun's gravitational field, the advance in the perihelion of mercury and the gravitational redshift of the light emitted from very massive objects. In all these cases the minor corrections to Newtonian gravity involved always *positively* curved distortions ($k > 0$) in the space-time. So the possibility of *negatively* curved distortions ($k < 0$) were probably not contemplated at this stage.

Only after Friedmann worked out systematically the set of all possible solutions (with $\Lambda = 0$) the possibility of a negatively curved space, involving anti-gravitational effects, was realized. Note that at the time the discovery of the CBR (and the possibility of cosmic

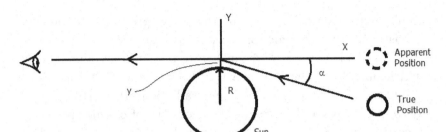

Fig. 6.1.   Deviation of light ray from a star when it passes near the Sun's surface: $\alpha \approx sGM/Rc^2$ where s is a dimensionless factor of order unity, G Newton's constant, M the Sun's mass, R the Sun's radius and c the speed of light.

radiation pressure effects) was yet to come many years into the future.

Let us consider[4] the bending of light rays by the Sun's gravitational field.

If a material particle or a photon with mass m moves in the vicinity of a body of large mass M, like the Sun, the first-order corrections to Newton's gravitational potential may be expected to be of the order of

$$E_p/mc^2 = (GMm/r)/mc^2 = GM/c^2r, \tag{6.3}$$

which is independent of m, the mass of the particle. Let us assume that the observed intervals of space and time, under the gravitational distortion due to M, are given, in a first approximation, by

$$\Delta x' \approx \left(1 - a\frac{GM}{c^2r}\right)\Delta x, \tag{6.4}$$

$$\Delta t' \approx \left(1 - b\frac{GM}{c^2r}\right)\Delta t, \tag{6.5}$$

where a and b are dimensionless constants of order unity.

The velocity of the moving particle, under these assumptions, would be given by

$$v' \approx \frac{\left(1 - a\frac{GM}{c^2r}\right)}{\left(1 - b\frac{GM}{c^2r}\right)}v = \left(1 - s\frac{GM}{c^2r}\right)v \tag{6.6}$$

where $s = a - b$ should be also of order unity. In particular, for the velocity of light under the influence of the strong gravitational field, one would have

$$c' = \left(1 - s\frac{GM}{c^2 r}\right) c. \tag{6.7}$$

To estimate the deviation of the light ray by the gravitational field of the Sun we set

$$d\alpha = \frac{dx' - dx}{dy} = \left(\frac{dc'}{dy}\right) dt = \frac{1}{c'} \left(\frac{dc'}{dy}\right) dx, \tag{6.8}$$

where c' is given by Eq. (6.7). Here

$$\frac{dc'}{dy} = s\frac{GM}{c^2 r}\frac{dr}{dy}, \tag{6.9}$$

and for a ray tangent to the Sun's surface $\left(y = 0, \ \frac{dr}{dy} = \frac{R}{r}\right)$

$$\frac{dc'}{dy} = \frac{2GMR}{cr^3}. \tag{6.10}$$

Therefore

$$\Delta\alpha \approx \frac{s\,GMR}{c^2} \int_{-\infty}^{\infty} \frac{dx}{r^3} = \frac{s\,GMR}{c^2}\left[\frac{2}{R^2}\right] = \frac{2s\,GM}{c^2 R}. \tag{6.11}$$

Substituting numerical data

$$G = 6.67 \times 10^{-8} \ (\text{cgs units})$$
$$M = 2 \times 10^{33} \ \text{g}$$
$$c = 3 \times 10^{10} \ \text{cm/s}$$
$$R = 6.96 \times 10^{10} \ \text{cm}$$

we get, assuming $s \approx 1$,

$$\Delta\alpha \approx 2.1 \times 10^{-6} \ \text{rad} \approx 0.43 \ \text{arcsecond},$$

which is a small but observable quantity.

But the most ambitious application of the General Theory of Relativity was made by Einstein himself when he applied it to the entire observable universe. And he concluded that, to avoid physical (not necessary mathematical) inconsistencies the universe must be finite.

Note that, as pointed out by Stanley L. Jaki,[5] Newton preferred to use the expression "an immense universe" instead of "an infinite" universe but, from his correspondence with Bentley, it appears that he argued in favor of a universe with infinite stars, overlooking the fact that a universe with infinite stars homogeneously distributed would give rise to infinite local gravitational forces pulling apart any material object in every direction.

In a closed universe (which would have a maximum finite space-time volume) extending a line element along a geodesic (the equivalent to a straight line in a Euclidean space) would end up coming back to the same point where it had begun. In an open universe (which has no boundaries, i.e., can grow indefinitely in space-time remaining finite) extending one line element along a geodesic would never come back to the point of departure. This is what would happen to a light ray propagating through either curved space. And also to material particles, or, for that matter, to very massive galaxies, moving in the respective local cosmic geometry.

In an open universe the particles, or galaxies, would not only move under the influence of the cosmic mass within the sphere interior to the point where the particle (or galaxy) was, *but* also, due to the peculiar geometry of the open space-time, they would notice a *counter-gravitational* force which could be in principle like a cosmic radiation pressure, like the one associated to the Cosmic Background Radiation at that particular point in space-time. As the cosmic radius R increases, the corresponding cosmic temperature T (associated with the CBR) decreases, and so does the cosmic radiation pressure, approaching zero but never reaching zero in a finite time.

The first one to take seriously the idea of an *open universe* ($k < 0$) was the Russian mathematician Alexander Friedmann[3] in 1922. The second one, independently, was the former World War I artillery officer (and later Catholic priest) Georges Lemaitre[6] in 1927. Lemaitre connected his open solutions to Einstein's equation with the *general expansion* of galaxies in the universe being observed at that time in the Mount Wilson Observatory by E. Hubble and M. L. Humason. He was the first to infer a primeval explosion of matter and energy out of a tremendously dense and hot *primeval atom*, introducing in this way what was latter called the Big Bang model of the origin of the universe.

References

1. S. Weinberg, *Gravitation and Cosmology* (John Wiley and Sons, New York, 1972).
2. S. Weinberg, *Cosmology* (Oxford University Press, Oxford, 2008).
3. A. Friedmann, *Z. Phys.* **10** (1922) 377.
4. M. Alonso and E. J. Finn, *Physics* (Addison-Wesley, Reading, Massachusetts, 1992).
5. S. L. Jaki, *The Road of Science and the Ways to God* (The Universe of Chicago Press, Chicago, 1978).
6. G. Lemaitre, *Ann. Soc. Sci. Brux.* **A47** (1927) 49.

Chapter 7

The Friedmann–Lemaitre Solutions

As we have anticipated in Chapter 1, Eq. (1.1), and discussed further in Chapter 6, Eq. (6.2), Einstein's cosmological equation is

$$\dot{R}^2 = \frac{8\pi}{3}G\rho R^2 - kc^2 + \frac{\Lambda}{3}c^2 R^2, \qquad (7.1)$$

where R is the cosmic radius, $\dot{R}$ its time derivative, G Newton's gravitational constant ($G = 6.67 \times 10^{-8}$ in cgs units), ρ the mass density (including matter mass and radiation mass), k the dimensionless space-time curvature, c the speed of light in vacuum ($c = 3 \times 10^{10}$ cm/s), and Λ the Einstein's cosmological constant.

We have mentioned previously that the term involving the cosmological constant becomes equivalent to the term involving the space-time curvature by indentifying $\Lambda = -3k^2/R^2$. So, assuming $k = 0$, $\Lambda \neq 0$ is fully equivalent to assuming $k = -|k| \neq 0$, $\Lambda = 0$. For a spherical universe of finite mass and open ($k < 0$) space-time geometry, Eq. (7.1) becomes

$$\dot{R}^2 = \frac{2GM}{R} + |k|c^2, \qquad (7.2)$$

where $M = \frac{4\pi}{3}R^3\rho$ is the total constant mass of the universe.

It is convenient to define a characteristic cosmic radius R_+ such that

$$\frac{2GM}{R_+} = |k|c^2, \quad R_+ = \frac{2GM}{|k|c^2}. \tag{7.3}$$

This radius is directly related to the cosmic Schwarzschild radius for a universe with mass M and sets the length scale for the universe.

Note that for $R \ll R_+$ the second term in Eq. (7.1) can be neglected and the universe's space-time geometry becomes indistinguishable from Euclidean ($k = 0$). For $R \gg R_+$ on the other hand the first term can be neglected, and the speed at which the cosmic radius grows approaches $\dot{R} \approx |k|^{1/2}c$. Note that if $M \to \infty$, $R_+ \to \infty$, and then the space-time geometry remains Euclidean at any $R < R_+$. Precisely because a universe with an infinite mass is irremediably paradoxical, cosmic space-time cannot be Euclidean.

Thus we are justified to assume that the universe has a finite mass and is open.

Lemaitre, considered the father of Big Bang cosmology, describes[1] the beginning of the universe in the following words:

> *The atomic world broke into fragments and each fragment in smaller bits. Assuming, for simplicity, that this fragmentation took place in bits of equal size, two hundred and sixty successive fragmentations must have taken place so as to achieve the fragmentation necessary to reach the fragmentation of our poor little atoms, so small that they can hardly be broken further. The world evolution can be compared to a just finished fireworks event: a few extinguishing red fires, ashes and smoke. Facing a cooling red hot coal, we see the slow end of the suns and try to imagine the vanishing brightness of the origin of the worlds.*

About twenty years later, in November 22, 1951, Pope Pius XII addressed the Pontificia Academia of Sciences, of which Lemaitre was then president, suggesting a certain amount of agreement between the original condition of the first matter of the universe according to some current scientific theories, and the august instant of the "Fiat Lux".

Fred Hoyle,[2] a staunch supporter of the Steady State Theory, then a dignified rival of the Big Bang, said "What kind of scientific theory is this that was conceived by a priest and endorsed by a pope?"

For years the Big Bang theory was treated with suspicion, especially because the age of the universe deduced by Lemaitre from Hubble's estimate of $\dot{R}/R$ was much less than the current Earth's age determined by means of radioactive dating. This was so because the cosmic distance scale used by Hubble was in error by a factor of ten.

Only after the discovery[3] of the cosmic background radiation by Penzias and Wilson the Big Bang theory, at first reluctantly, then enthusiastically begun to be accepted by the scientific community at large.

Let us go back to Einstein's cosmological compact Friedmann–Lemaitre solutions in terms of R_+. They are obtained as follows: Eq. (7.1) for an open universe is rewritten as

$$\dot{R}^2 = R^{-1/2} \left\{ \frac{8\pi G}{3} \rho R^3 + c^2 |k| R \right\}^{1/2} \tag{7.4}$$

which can be integrated as

$$\int dt = \int \frac{R^{1/2}}{\{2GM + c^2|k|R\}^{1/2}} dR. \tag{7.5}$$

Using the change of variable

$$x^2 = c^2 |k| R \tag{7.6}$$

it can be rewritten as

$$\int \frac{x^2}{(a^2 + x^2)^{1/2}} dx \tag{7.7}$$

which can be found in tables, resulting in

$$t = \frac{R_+}{c|k|^{1/2}} [\sinh(y)\cosh(y) - y], \tag{7.8}$$

where $y \equiv \sinh^{-1}(R/R_+)^{1/2}$, which implies

$$R = R_+ \sinh^2 y. \tag{7.9}$$

From the compact parametric solutions $t(y)$, $R(y)$, in terms of $R_+ = 2GM/|k|c^2$, we can get easily

$$\dot{R} = |k|^{1/2}c/\tanh y, \tag{7.10}$$

$$\ddot{R} = -\frac{1}{2}\frac{|k|^{1/2}c}{R_+}\Big/\sinh^4 y, \tag{7.11}$$

$$H = \dot{R}/R = \frac{|k|^{1/2}c}{R_+}\cosh y \Big/ \sinh^3 y, \tag{7.12}$$

$$\rho/\rho_c \equiv \Omega = 1/\cosh^2 y = 1 - \tanh^2 y, \tag{7.13}$$

$$-\ddot{R}R/\dot{R}^2 \equiv q = \frac{1}{2|k|^{1/2}}(1/\cosh^2 y). \tag{7.14}$$

In particular, the dimensionless product Ht is given by

$$Ht = \frac{[\sinh(y)\cosh(y) - y]\cosh(y)}{\sinh^3(y)} \tag{7.15}$$

and therefore $2/3 \leq Ht \leq 1$ for y going from 0 to $\gg 1$.

It may be noted that using the present observational data[4] for $H_o = 67\,\text{km/sMpc} = 2.18 \times 10^{-18}\,\text{s}^{-1}$ (within 5.6%) and to $13.7\,\text{Gyrs} = 4.32 \times 10^{17}\,\text{s}$ (within 1.5%) the product

$$H_o t_o = 0.942 \pm 0.065 \tag{7.16}$$

is incompatible with a closed universe ($k > 0$), which implies $2/3 \geq Ht \geq 0$.

On the other hand, the dimensionless ratio $\Omega = \rho/\rho_c$, where $\rho = \rho_m + \rho_r$ is the sum of the matter mass density and the radiation mass density, and where $\rho_c = 3H^2/8\pi G$ is the critical mass density, is given by

$$\Omega = \Omega_m + \Omega_r = \frac{1}{\cosh^2(y)}1 - \tanh^2(y), \tag{7.17}$$

and therefore $1 \leq \Omega \leq 0$ for y going from 0 to $\gg 1$.

Note that the present observational values for H_o and t_o fix the present value of y_o through

$$H_o t_o = \frac{[\sinh(y_o)\cosh(y_o) - y_o]\cosh(y_o)}{\sinh^3(y_o)} = 0.942 \pm 0.065 \tag{7.18}$$

resulting in

$$y_o = \sinh^{-1}(R_o/R_+) = \sinh^{-1}(T_+/T_o) = 2.243 \pm 0.157 \quad (7.19)$$

where the equation of state $RT = R_oT_o = $ constant (the present transparent phase of the universe) has been used. Taking into account that $R = R_+$ (Schwarzschild radius), $\dot{R}_+ = |k|c^2/\tanh(y) \approx c$, where $\sinh^{-1}(y_+) = 1$, resulting in $|k| = 1/2$, and Eq. (7.8) for $y = y_o = 2.243$, $t(y) = t(y_o)$ results in

$$R_+ = 4.588 \times 10^{26} \text{ cm}$$
$$(R_o = R_+ \sinh^2 y_o = 9.96 \times 10^{27} \text{ cm}),$$

and, correspondingly, that

$$T_+ = (R_o/R_+)T_o = (21.69)2.726 = 59.2 \,\text{K},$$

the set of principal cosmic numbers being sufficiently specified. Using $y_o = 2.243$, the dimensionless ratio

$$\Omega_o = \Omega_{mo} + \Omega_{ro} = 1 - \tanh^2 y_o = 0.044 \quad (7.20)$$

is obtained. This is again incompatible with a closed universe $(k > 0)$, which implies $1 \leq \Omega$.

Knowing numerically R_+ and $|k|$ the finite total matter mass of the universe can be evaluated as being

$$M = \frac{R_+c^2}{2G|k|^{1/2}} = 1.547 \times 10^{54} \text{ g.} \quad (7.21)$$

Dividing Einstein's cosmological equation (7.1) with $k < 0$, $\Lambda = 0$, by $\frac{8\pi}{3}GR^2$ we get

$$\frac{3(\dot{R}/R)^2}{8\pi G} = \rho + \frac{3(\dot{R}/R)^2}{8\pi G}|k|\left(\frac{c}{\dot{R}}\right)^2, \quad (7.22)$$

and, taking into account that $\frac{3(\dot{R}/R)^2}{8\pi G} = \rho_c$ (the critical density), we get

$$1 = \left(\frac{\rho}{\rho_c}\right) + |k|\left(\frac{c}{\dot{R}}\right)^2 = (\Omega_m + \Omega_r) + \Omega_k, \quad (7.23)$$

where we have defined Ω_k as $\Omega_k = |k|(c/\dot{R})^2$. If we identify the left-hand side of Eq. (7.1) with the *kinetic energy* and the first term in the right-hand side Eq. (7.1) with the *gravitational potential energy*, we could identify the second term Ω_k as *the space-time curvature potential energy*. Then

$$\Omega = \frac{1}{\cosh^2 y} = 1 - \tanh^2 y, \tag{7.24}$$

$$\Omega_k = 1 - \Omega = \tanh^2 y. \tag{7.25}$$

To separate the mass matter and mass radiation contribution in Ω we make use of the equation of state for the transparent phase of the universe's expansion ($R \geq R_{af}$, $T \leq T_{af}$), where R_{af} and T_{af} are the radius and CBR temperature at decoupling ($\rho_{maf} = \rho_{raf}$, atom formation)

$$RT = R_{af}T_{af} = \text{constant}, \tag{7.26}$$

and also of the fact that

$$\rho_m(R) = \rho_{maf} \left(\frac{R_{af}}{R}\right)^3, \left[\rho_m(R) = \frac{M}{\frac{4\pi}{3}R^3}\right], \tag{7.27}$$

$$\rho_r(R) = \rho_{raf} \left(\frac{T}{T_{af}}\right)^4, \left[\rho_r(T) = \sigma T^4\right], \tag{7.28}$$

to get

$$\frac{\rho_r}{\rho_m} = \left(\frac{RT}{R_{af}T_{af}}\right)^3 \frac{T}{T_{af}} = \frac{T}{T_{af}}. \tag{7.29}$$

Then

$$\Omega_m = \Omega \bigg/ \left(1 + \frac{T}{T_{af}}\right), \tag{7.30}$$

$$\Omega_r = \Omega \left(\frac{T}{T_{af}}\right) \bigg/ \left(1 + \frac{T}{T_{af}}\right). \tag{7.31}$$

Here T (decoupling) $= T_{af} = 2968\,\text{K}$ and $T_+ = 59.2\,\text{K}$ (as pointed out above for $y_o = 2.243$).[5]

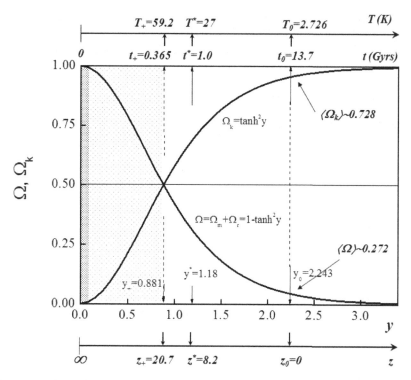

Fig. 7.1. Contributions to the density parameter Ω_m (matter), Ω_r (radiation), Ω_k (space-time curvature) as a function of y (dimensionless cosmic parameter), z (redshift), t (time) and T (CBR temperature).

Figure 7.1 displays the evolution of Ω (matter plus radiation), and Ω_k (space-time curvature) as a function of $T(K)$, $t(\text{Gyrs})$, y (dimensionless cosmic parameter), and z (redshift) from $y = 0$ to $y \approx 3.5$, which spans the early plasma event at which atoms form ($T_{af} = 2968\,\text{K}$), the moment ($\Omega = \Omega_k$) at which stars and galaxies begin to form ($T_+ = 59.2\,\text{K}$), and the present epoch ($T = 2.726\,\text{K}$).

The universe as we see it now encompasses from protogalaxies and protostars formed at about $T^* \approx 27\,\text{K}$, $t^* \approx 1\,\text{Gyr}$, $y^* \approx 1.18$, $z^* \approx 8.2$ to galaxies and stars formed more recently as we see them now at $T_o \approx 2.726\,\text{K}$, $t_o \approx 13.7\,\text{Gyrs}$, $y_o \approx 2.243$, $z_o \approx 0$.

References

1. G. Gammow, in *The Creation of the Universe* (Viking Press, New York, 1952).
2. J. C. Mather and J. Boslough, *The Very First Light* (Penguin Books, London, 1996), p. 48.
3. A. A. Penzias and R. W. Wilson, *Astrophys. Journal*, **142** (1965) 419.
4. S. Weinberg, *Cosmology* (Oxford University Press, Oxford, 2008).
5. J. A. Gonzalo, *Inflationary Cosmology Revisited* (World Scientific, Singapore, 2005).

Chapter 8

The Role of Radiation Pressure

Alan Lightman and Roberta Brawer in *Origins*[1] (a set of interviews with leading cosmologists giving their perspective in 1990) suggest the year 1965 — the year in which Arno Penzias and Robert Wilson reported their discovery of the *cosmic background radiation* (CBR) — as the starting year of scientific cosmology.

This suggestion may be a little exaggerated. But, as Ralph Alpher and Robert Herman — the predictors of the CBR in 1948 — note in their latest book,[2] the year 1965 was a pivotal year for scientific cosmology. The importance of the cosmic radiation, which had been more or less overlooked by Einstein, Friedmann, Lemaitre, Eddington, Hubble and other pioneer cosmologists, would be certainly decisive in bringing to maturity modern scientific cosmology.

In 1964, satellite communications was entering a period of rapid development. Researchers at leading laboratories were improving their equipment to detect clean signals for commercial exploitation of the new technology. That year A. Penzias and R. Wilson, at Bell Labs, were developing a microwave receiver with a very low noise figure, with the aim of using it as a receiver for satellite communications. As it is well known, after eliminating carefully all conceivable sources of unwanted noise in their antenna, they were left with a persistent strange signal. It was the background radiation, which in the

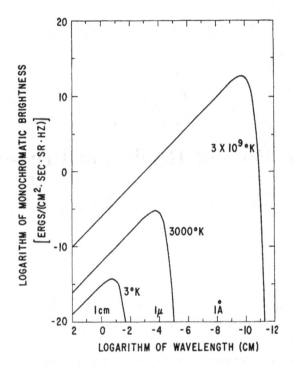

Fig. 8.1.   Planck's blackbody radiation spectrum giving the relative (monochromatic) brightness in ergs/cm^2·sec as a function of the wavelength in cm. The CBR observed by Penzias and Wilson in 1965 at 7.3 cm corresponded to a characteristic temperature of about 3.5 K.

present cosmic epoch is at a wavelength of 7.3 cm in the microwave region of the electromagnetic spectrum. The persistent background radiation appeared to be unpolarized and isotropic, independent of the direction of observation, and with an equivalent temperature of about 3.5 K if the source was assumed to be blackbody, as later confirmed. The blackbody character of the radiation, as shown by Max Planck in 1900, was completely specified by a single number: the value of the equivalent temperature. (See Fig. 8.1.)

In their 1965 paper[3] Penzias and Wilson reported that, at first, they did not know they were making a fundamental cosmological observation but that was precisely what it was. Given the blackbody character of the CBR, one measurement at a single wavelength was sufficient to characterize the spectrum in the full spectral range.

Let us describe briefly the role played by the cosmic background radiation through the expansion.

In its early *plasma phase* (at $T > T_{\text{decoupling}} \approx T_{\text{atom formation}}$) the universe was opaque, and matter and radiation had to expand in unison. In fact radiation pressure was pushing matter away radially (i.e., material particles, after baryon formation, electron formation and primordial nucleosynthesis). In this earlier *plasma* phase

$$\rho = \frac{M}{\frac{4\pi}{3}R^3} \quad \text{(matter mass density)}, \tag{8.1}$$

and

$$\rho = \frac{\sigma T^4}{c^2} \quad \text{(radiation mass density)}, \tag{8.2}$$

according to the Stefan–Boltzmann law for blackbody radiation. The cosmic equation of state before decoupling (atom formation) was therefore

$$\rho_r/\rho_m = \rho_{\text{raf}}/\rho_{\text{maf}} \to T^4 R^3 = T_{\text{af}}^4 R_{\text{af}}^3 = \text{constant} \ (T \leq T_{\text{af}}). \tag{8.3}$$

Then, at $t = t_{\text{af}}$ the cosmic temperature had decreased sufficiently to let electrons and protons (or ^{4}He nuclei) form hydrogen and helium atoms. The universe became transparent, and radiation propagated much more freely; because in a transparent universe the total number of photons is conserved, as is the total number of baryons, we have

$$n_r/n_b = \left(\frac{\sigma T^4}{2.8\,k_B T}\right) \Big/ \left(\frac{1}{m_b}\frac{M}{4\pi R^3/3}\right)$$

$$= \text{constant} \times T^3 R^3 = \text{constant} \tag{8.4}$$

resulting in a different equation of state

$$T^3 R^3 = T_{\text{af}}^3 R_{\text{af}}^3 = \text{constant} \ (T \geq T_{\text{af}}). \tag{8.5}$$

Note that the ratio ρ_r/ρ_m is no longer constant (and equal to unity) at $t \geq t_{\text{af}}$ but becomes instead

$$\rho_r/\rho_m = (\rho_{\text{raf}}/\rho_{\text{maf}}) \cdot (TR/T_{\text{af}}R_{\text{af}})^3 \cdot (T/T_{\text{af}}) = T/T_{\text{af}}. \tag{8.6}$$

We can estimate the *ratio* of radiation pressure

$$p_r = \frac{1}{3}\sigma T^4$$

to matter pressure

$$p_m = \frac{n k_B T}{V} = \left(\frac{n m_b}{V}\right) \frac{k_B T}{m_b} = \rho_m \frac{k_B T}{m_b},$$

where n is the number of baryons per unit volume and m_b the baryon mass. Then

$$\frac{p_r}{p_m} = \frac{\frac{1}{3}\rho_r c^2}{\rho_m \frac{k_B T}{m_b}} = \frac{1}{3}\left(\frac{T}{T_{af}}\right)\frac{m_b c^2}{k_B T} = \frac{1}{3}\frac{m_b c^2}{k_B T_{af}} \approx 9.35 \times 10^8. \quad (8.7)$$

In other words, the *radiation pressure* in the present transparent phase of cosmic expansion is much larger than the *matter pressure*, itself larger than the gravitational attraction. This is in consonance with the *open* $(k < 0)$ space-time character of our universe.

We can make a direct comparison among the ways k affects cosmic space-time for the three cases of an *open* $(k < 0)$, a Euclidean or *flat* $(k = 0)$, and a *closed* $(k > 0)$ universe. For an *open* universe, as we have seen

$$R = R_+ \sinh^2 y, \quad t = \frac{R_+}{|k|^{1/2}c}\{\sinh y \cosh y - y\}, \quad (8.8)$$

where

$$\sinh y \approx y + \frac{y^3}{3!} + \frac{y^5}{5!}\ldots, \quad \cosh y \approx 1 + \frac{y^2}{2!} + \frac{y^4}{4!} + \frac{y^6}{6!}\ldots.$$

For a *flat* universe, especially for $y \ll 1$, we can interpolate

$$R = R_+ y^2, \quad t = \frac{2}{3}\frac{R_+}{c}y^3. \quad (8.9)$$

For a *closed* universe,

$$R = R_+ \sin^2 y, \quad t = \frac{R_+}{|k|^{1/2}c}\{y - \sin y \cos y\}, \quad (8.10)$$

where

$$\sin y \approx y - \frac{y^3}{3!} + \frac{y^5}{5!}\ldots, \quad \cos y \approx 1 - \frac{y^2}{2!} + \frac{y^4}{4!} - \frac{y^6}{6!}\ldots.$$

Therefore, in an *open* $(k < 0)$ universe, space and time are both dilated with respect to the *flat* $(k = 0)$ universe:

$$\frac{[R(y)]_{k<0}}{[R(y)]_{k=0}} > 1, \quad \frac{[t(y)]_{k<0}}{[t(y)]_{k=0}} > 1.$$

And in a *closed* ($k > 0$) universe, space and time are both *contracted* with respect to the flat ($k = 0$) universe

$$\frac{[R(y)]_{k>0}}{[R(y)]_{k=0}} < 1, \quad \frac{[t(y)]_{k>0}}{[t(y)]_{k=0}} < 1.$$

The growth speed $\dot{R}$ of the cosmic radius at $y \ll 1$ is therefore given by

$$\dot{R} = \frac{dR/dy}{dt/dy} = c(|k|^{1/2}/y)[1 + sy^2 + \cdots] \text{ for } k < 0,$$

$$\dot{R} = \frac{dR/dy}{dt/dy} = c(1/y) \quad \text{for } k = 0, \text{ and}$$

$$\dot{R} = \frac{dR/dy}{dt/dy} = c(|k|^{1/2}/y)[1 - sy^2 + \cdots] \text{ for } k > 0,$$

with s a numerical factor of order unitary. Note that at $y \ll 1$, $\dot{R} \gg c$.

Can then $k < 0$, $k = 0$ and $k > 0$ be given physical meaning? The inequality $k < 0$ can tentatively be associated with radiation pressure countering gravity due to the cosmic background radiation; $k = 0$ with an absolutely void space-time; and $k > 0$ with homogeneously distributed cosmic dust or cosmic matter of unknown composition, reinforcing gravity.

References

1. A. Lightman and R. Brawer, *Origins: The Lives and Worlds of Modern Cosmologists* (Harvard University Press, Cambridge, MA, 1990).
2. R. A. Alpher and R. Herman, *Genesis of the Big Bang* (Oxford University Press, Oxford, 2001).
3. A. A. Penzias and R. W. Wilson, *Ap. J.* **142** (1965) 419.

Chapter 9

The Einstein–Lemaitre Correspondence

The cosmological expansion is, because its scale, the most grandiose phenomenon in nature. As we have seen, it was predicted theoretically[1] by the Russian mathematicians Alexander Alexandrovich Friedmann (1888–1925) in 1921 and, independently, by the Belgian astrophysicist and Catholic priest Georges Edouard Lemaitre (1894–1966) in 1927.

In October 1927, at Brussels, during the Solvay Conference on Physics (October 24–29) Albert Einstein met a relatively young Lemaitre (then 33), a Catholic priest wearing Roman collar, who ten years before had been an artillery officer in the Belgium Army at World War One. At the time of their encounter, Einstein probably did remember Lemaitre's paper[2] on solutions of his relativistic cosmological equations. This paper may have made him think of the article by Friedmann, the Russian mathematician who had obtained solutions for his cosmological equation which implied an expanding universe.

Einstein was then a little rude with the young cosmologist and Catholic priest: "Your calculations are correct, but your physics is abominable."[3] This is the same reaction he had to the work of the

Russian mathematician in 1922 (Friedmann had died in 1925; his disciple George Gamow, who would develop twenty years later Friedmann's model of the expanding universe, popularized the Big Bang model).

By 1927 Einstein was beginning to consider seriously the mathematical correctness of Friedmann and Lemaitre's expanding solutions of his cosmological equations, but he was still unconvinced. By then, however Lemaitre had gone much further than Friedmann. After spending a year in Cambridge, UK, with Arthur S. Eddington, the first champion of Einstein's theory of relativity in England (who became a world celebrity after the African eclipse expedition which detected the bending of starlight by the Sun), Lemaitre was encouraged by Eddington to travel to Cambridge, Massachusetts, for another year, to work with Harlow Shapley, who had been mapping the Milky Way using the 60-inch telescope of Mount Wilson, California.

There were already strong indications that cosmic nebulae far away were receding from us in space. This was taken at least as preliminary experimental evidence of an expanding cosmos. As noted by Farrel,[3] Lemaitre accompanied Einstein in Brussels to Piccard's lab and brought up the subject of recent measurements by American astronomers on spiral nebulae, suggesting that the redshifts in their spectra were indicative of a general expansion of galaxies in the universe. Two years later Einstein did begin to consider the possibility that the overall expansion of the universe was real. Within about two years, he had begun to accept the expansion publicly.

According to Helge Kragh[4]

In his (Lemaitre's) 1933 address (to the American National Academy of Sciences) he suggested an interesting interpretation of the (cosmological) constant, namely, that it may be understood as a negative vacuum density: "Everything happens as though the energy in vacuo would be different from zero", he wrote, referring to general relativity applied to regions of space of extremely low density. He then argued that in order to avoid a nonrelativistic vacuum or ether — a medium in which absolute motion is detectable — a negative pressure $p = -\rho c^2$ must be introduced, the vacuum density being related to the cosmological constant by $\rho = \Lambda c^2 / 4\pi G$. This negative *pressure is responsible for the exponential (de Sitter) expansion of Lemaitre's universe during its last phase, and it contributes a repulsive cosmic force of $\Lambda c^2 r/3$.*

(Here ρ = density, c = speed of light, Λ = cosmological constant, G = Newton's constant of gravity).

Note that, as pointed out in the previous chapter the *negative vacuum density* of Lemaitre can be advantageously substituted by *the radiation pressure due to the CBR* (cosmic background radiation) associated either to the negative space-time curvature ($k < 0$), to the cosmological constant ($\Lambda \neq 0$) or to both.

After their first meeting in Brussels in 1927, Lemaitre met Einstein again several times and corresponded for many years afterwards. Einstein's correspondences with Lemaitre on the famous cosmological constant are given in the first page of the letter which he[5] wrote to Lemaitre in September 28, 1947:

> *Dear Professor Lemaitre,*
>
> *I thank you very much for your kind letter of July 30. In the meantime I received from Professor Schlipp your interesting paper for his book. I doubt that anybody has so carefully studied the cosmological implications of the theory of relativity as you have. I can also understand that in the shortness of T_o (present "age" of the universe, NT) there exists a reason to try bold extrapolations and hypothesis to avoid contradiction with facts. It is true that the introduction of the Λ term offers a possibility, it may even be that it is the right one.*
>
> *Since I have introduced this term I had always a bad conscience. But at that time I should see no other possibility to deal with the fact of the existence of a finite mean density of matter. I found it very ugly indeed that the field law of gravitation should be composed of two logically independent terms which are connected by addition. About the justification of such feelings concerning logical simplicity it is difficult to argue. I cannot help to feel it strongly and I am unable to believe that such an ugly thing should be realized in nature.*

In his book *The Primeval Atom*,[6] Lemaitre says:

> *We can compare space-time to an open, conic cup... The bottom of the cup is the origin of atomic disintegration (the disintegration of the "primeval atom", NT); it is the first instant at the bottom*

of space-time, the now which has no yesterday (*emphasis added*)
because, yesterday, there was no space.

In my opinion, Einstein was right in dismissing his cosmological constant, but, given the effective equivalence between the *space-time curvature term with* $\Lambda > 0$, Lemaitre was perhaps not completely wrong.

References

1. M. Heller and A. Chernin, *Los Origenes de la Cosmologia: Friedmann y Lemaitre,* Translated to Spanish from the Russian edition (Editorial URSS-Libros de ciencia, Moscú, 1991).
2. G. Lemaitre, *Ann. Soc. Sci. Brux. A* **47** (1927) 49, p. 12.
3. J. Farrell, *The Day without Yesterday* (Tunder's Mouth Press: New York, 2005), p. 10.
4. H. Kragh, *Cosmology and Controversy: The Historical Development of Two Theories of the Universe* (Princeton University Press, Princeton, N.J., 1996), p. 53.
5. Einstein's letter to Lemaitre in September 1947 (Albert Einstein Archives, Jewish National and University Library); reproduced in Ref. 4, p. 143.
6. G. Lemaitre, *The Primeval Atom: A Hypothesis of the Origin of the Universe,* Translated by B. H. Korff and S. A. Korff (D. Van Nostrand, New York, 1950).

Chapter 10

The Universal Constants

Heisenberg's uncertainty principle has an intrinsic connection with the universal constant $\hbar$, Planck's quantum of action divided by 2π:

$$\hbar = 1.05 \times 10^{-27} \text{ erg} \cdot \text{sec}. \tag{10.1}$$

On April 11, 2000, at 12:30 pm an academic session was held at the main Conference Room of the Faculty of Science, Universidad Autónoma de Madrid (UAM) to celebrate the 1st centennial of **Planck's quantum of action**. That session was opened by Professor Rodolfo Miranda, Vice Chancellor for Research, on behalf of Professor Emilio Crespo, Vice Chancellor for Cultural Activities. He pointed out that Planck's discovery would change the way scientists perceive physical reality in the future by his preparing the way for the uncertainty principle. The next speaker, Professor Agustín Gárate, Vice Dean of Professorate of the Faculty of Science, said that as teachers and students of physical sciences (especially physics and chemistry) we know very well the tremendous impact of Planck's work in the twentieth century. Planck's work and Planck's constant (h) led very soon to the understanding of the photoelectric effect (Einstein), the specific heat of solids (Einstein and Debye) and the atomic spectra of the hydrogen atom (Bohr). Then I introduced the Invited Lecturer for this occasion, Professor Stanley L. Jaki,

Distinguished Professor at Seton Hall University, South Orange, New Jersey, Templeton Prize, 1987. Professor Jaki had visited UAM in 1992, when he gave a very interesting lecture on "Is there such a thing as a last word in physics?"; and then in 1997, when he spoke about "Lucky coincidences at the Earth–Moon system and their relevance for Drake's equation". At the Academic Session he entitled his lecture "Numbers decide or Planck's constant and some constants of philosophy", quoting Planck's words at his Nobel Lecture, June 2, 1920, at which he tried to give in his own words the story of the origin of quantum theory in broad outline, its development up to that time, and its significance for physics.

In his first illustration, Professor Jaki showed a picture with Planck's tombstone at the cemetery of Gottingen in which Planck's effigy and his name appeared together with his quantum's numerical expression,

$$\hbar = 6.67 \times 10^{-27} \text{ erg} \cdot \text{sec}.$$

Then he proceeded to do a detailed and masterful analysis of the ups and downs of Planck's discovery of the "quantum of action" as a result of his laborious interpretation of the "black body" emission in the whole range of wavelengths. He showed convincingly the validity of Planck's summary of his epoch-making discovery in just two words "**Numbers decide**". He showed[1] first Wien's formula which described very well black body radiation at low wavelengths

$$\phi_\lambda = \frac{C}{\lambda^5} e^{-c/\lambda\theta} \text{ (Wien, 1896)} \tag{10.2}$$

where C and c are constants, λ is the radiation wavelength and θ the absolute temperature. And then he showed Planck's formulae, in quick succession, leading to the description of black body radiation at any wavelength by

$$E = \frac{8\pi ch}{\lambda^5} \frac{1}{e^{ch/k\lambda\theta} - 1}. \text{ (Planck, Jan 7, 1901)} \tag{10.3}$$

The expression "black body" or "black cavity" radiation was yet to come into wide use. Since those black body cavities give a minimum light at a given temperature, they were used as instruments of calibration for light sources (including light bulbs) which at that time were coming into wide use all over the place.

Wien's derivation of his displacement law (1897), which registered the displacement towards lower wavelengths of the emission maxima, called for widespread attention, and his work was quickly translated into English and published in the *Philosophical Magazine*, the most prestigious British journal for physics at the time. Wien had relied for his derivation on the Stephan–Boltzmann law.

In November 1899 Lummer and Pringsheim, and then Rubens, had published new results showing that when the temperature of the black body emitter was increased, there appeared deviations from Wien's law (7.2) at high wavelengths. In October 7, 1900, Rubens visited Planck and in the evening of that day Planck was quick to write to Rubens telling him that he had just obtained a new law which agreed very well with the new data. A few days later, on October 19, Planck presented his new formula to the *Physikalische Gesellschaft* including all recent data and stressing the point that the **numbers** were **decisive** in confirming what he saw as a **fundamental law of nature**.

Max Born, recalling a conversation[2] with Lummer and Pringsheim, wrote in 1906: "Although Planck's formula was in the center of discussion, one was nevertheless inclined to view Planck's proposal of **quantum-like oscillator energy** only as a provisional working hypothesis, and Einstein's light quanta were not taken seriously."

On February 27, 1909, Planck wrote to Wien (a good friend from his youth years) dismissing the radical inference of Einstein about the propagation of light in the form of **bundles** or **quanta** of energy.

On another occasion Planck recalled Wien's hunting skills as well as his skills as a physicist[3]: "There are but few physicists who mastered so equally as Willy Wien did the experimental and the theoretical parts of their science, and in the future it will happen on even fewer occasions that one and the same researcher would make so different discoveries as the displacement law of blackbody radiation and the nature of cathode rays."

Even in 1910 Planck wrote: "The introduction of the quantum of action h into the theory should be done as **conservatively** as possible, i.e., alterations should only be made that have shown themselves to be absolutely necessary." When in 1911 Wien received the Nobel Prize in Physics, no reference was made to the fact that Wien's formula had been shown to be incorrect in the infrared range. In his acceptance speech Wien both praised and criticized Planck, and suggested that one should look into Sommerfeld's interpretation of h to get a physical meaning. At that time Bohr's atom model was still not published.

In a paper presented by Planck at the Solvay Conference, Brussels, October 1911, where all the leading "progressive" physicists were present, he insisted that his hypothesis was not an "Energie-hypothese" but a "Wirkungshypothese".[4] In any case Planck brought to a conclusion his speech on that occasion as follows:[5]

> "The largest part of the work is still to be done. Surely some death flowers will keep falling from the tree of knowledge. But the beginning has already been made. The hypothesis of the quantum shall not disappear from the world. The laws of heat radiation guarantee this. And I do not think to go too far when I state my opinion that through that hypothesis the foundation is laid for the construction of a theory which is destined to cast in a new light the fast and delicate moving events of the molecular world."

The concept of universal constant must be traced back to Planck's theoretical work on black body radiation.[6] His quantum theory was successful to account for:

(a) **The spectral distribution** of black body radiation for all frequencies

$$W_T(\omega)\mathrm{d}\omega = \frac{\hbar}{\pi}\frac{\omega^3}{c^3}\frac{1}{e^{\hbar\omega/k_BT}-1}\mathrm{d}\omega, \qquad (10.4)$$

equivalent to Eq. (10.3), where W_T is the thermal radiation energy density per unit volume and per unit frequency interval, T the absolute temperature, ω the angular frequency, $\hbar = h/2\pi$, and k_B = Boltzmann's constant.

(b) **The Wien's displacement law**, according to which the maximum of the emitted radiation for a given equilibrium temperature T occurs at a certain ω_{max} such that it grows with T as

$$\hbar\omega_{max} \cong 2.8 k_B T; \tag{10.5}$$

and

(c) **The Stephan–Boltzmann law**, which gives the total amount of emitted radiation per unit volume in the whole frequency range from zero to infinity, as

$$\int_0^\infty W_T(\omega)d\omega = \left[\frac{\pi^2}{15}(\hbar c)^{-3}k_B^4\right]T^4 \equiv \sigma T^4. \tag{10.6}$$

These three relations allowed Planck to determine the set of **universal constants**

$$\hbar = h/2\pi = 1.05 \times 10^{-27} \text{ erg} \cdot \text{s}, \tag{10.7}$$

$$k_B = 1.38 \times 10^{-16} \text{ erg/K}, \tag{10.8}$$

$$c = 3 \times 10^{10} \text{ cm/s}. \tag{10.9}$$

These constants, together with Newton's gravitational constant,

$$G = 6.67 \times 10^{-8} \text{ cm} \cdot \text{g}^{-1} \cdot \text{s}^{-2},$$

allowed Planck to get a set of units[7] for mass, length, time (and temperature) which are independent of specific bodies and substances and necessarily keep their meanings for all times and for all cultures... and can be designated as "natural units".

References

1. Julio A. Gonzalo (coordinator), *Planck's Constant: 1900–2000* (Universidad Autónoma de Madrid: Servicio de Publicaciones, 2001), pp. 109–110.
2. Max Born, *Physik im Wandel meiner Zeit*, 4th Ed. (F. Vieweg, Braunsweig, 1966), p. 244.
3. Max Planck, "Dem Andenken and W. Wien", in K. Wien (ed.), *Wilhelm Wien: Aus dem Leben und Wirken eines Physikers* (Johan A. Barth, Leipzig, 1930), pp. 139–140.
4. Stanley L. Jaki, *The Road of Science and the Ways to God* (The University of Chicago Press, 1978); *Abhandlungen*, Vol. 2, p. 285.
5. *Ibid.*; *Abhandlungen*, Vol. 3, p. 64.
6. R. Loudon, *The Quantum Theory of Light*, 2nd Ed. (Clarendon Press, Oxford, 1983), Chap. 1.
7. Max Planck, *Wissenschflitche Selbsbiographie* (1948), p. 374.

Chapter 11

Rigorous Solutions of Einstein's Cosmological Equation

Manuel Alfonseca and myself submitted recently to arXiv.org our work "Comment on the 1% concordance Hubble constant". In it a summary of the solutions of Einstein's cosmological equations for an **open Friedmann–Lemaitre universe** and a flat **Lambda Cold Dark Matter universe** model are examined.

C. L. Bennet *et al.* have recently observed that the accurate determination of the Hubble constant $(\dot{R}_0/R_0)$ for $z \cong 0$ has been, and is, a central goal in observational astrophysics. After a careful analysis of the existing data they conclude that $H_0 = 69.6 \pm 0.7 \, \mathrm{kms}^{-1}\mathrm{Mpc}^{-1}$. This is almost compatible with $H_0 = 73.8 \pm 2.4 \, \mathrm{kms}^{-1}\mathrm{Mpc}^{-1}$ as recently evaluated by Riess *et al.* In this comment we note that the difference is significant at least in one respect: assuming $t_0 = 13.7$ Gyrs ($\pm 0.5\%$) the value given by C. L. Bennet *et al.* results in $H_0 t_0 = 0.975 < 1$, which is compatible at least marginally with an open universe ($k < 0$, $\Lambda = 0$); while the value given by Reiss *et al.*, $H_0 t_0 = 1.034 > 1$, is not. NASA's James Webb Space Telescope, to be launched in 2017, is expected to determine H_0 with an accuracy better than 1%.

It is well known that Hubble's original attempt (1929) to make a quantitative evaluation of the velocity–distance ratio for distant galaxies involved serious systematic errors. For a long time afterwards, Sandage, Hubble's successor at Mount Wilson Telescope favored a Hubble constant value of $H_0 \cong 50$ kms^{-1}Mpc^{-1}, while Vaucouleurs, another distinguished astronomer, favored $H_0 \cong 100$ kms^{-1}Mpc^{-1}. NASA's Hubble Space Telescope produced $H_0 \cong 72 \pm 8$ kms^{-1}Mpc^{-1} (Freedman *et al.*, 2001).[1] More recently in 2014, Riess *et al.*[2] have given $H_0 \cong 73.8 \pm 2.4$ kms^{-1}Mpc^{-1}, which in principle reduces the uncertainties to 3%. The problem is by no means settled and the quoted uncertainties may be a little optimistic.

Bennet *et al.*[3] have published a careful re-examination of the self-consistency of H_0 measurements as given by the WMAP and Planck satellites and some ground-based telescopes. The best fit, obtained after a meticulous evaluation, gives $H_0 \cong 69.6 \pm 0.7$ kms^{-1}Mpc^{-1}. Hopefully NASA's James Webb Space Telescope, to be launched in 2017, will give a vastly improved value for H_0, both for the central value and for the reduced uncertainty.

On the other hand, as pointed out long ago by Beatriz Tinsley in 1977,[4] the dimensionless product $H_0 t_0$ can be especially advantageous to characterize quantitatively the solutions of Einstein's cosmological equations. Using the analytic solutions of Einstein's equations for $\Lambda > 0$ (Lambda Cold Dark Matter model with $k = 0$) and for $k < 0$ (open Friedmann–Lemaitre model with $\Lambda = 0$) it is easy to check that[5]

$$\frac{2}{3} \leq H_0 t_0 \leq \infty \quad \text{(LCDM model)}, \tag{11.1}$$

and

$$\frac{2}{3} \leq H_0 t_0 \leq 1 \quad \text{(OFL model)}. \tag{11.2}$$

Here the numerical value depends principally on H_0, since t_0 was determined with magnificent accuracy by WMAP in 2003 and was confirmed by **Planck Satellite** in 2013 as $t_0 = 13.7 \pm 0.1$ Gyrs.

It may be instructive to take a look at characteristic cosmological parameters such as $\Omega = \Omega_m + \Omega_r$ (matter mass density plus radiation mass density divided by the critical density) or z_{Sch}

(the hypothetical maximum redshift corresponding to $R = R_{\text{Sch}}$ for the universe, obviously greater than $(z_{\text{obs}})_{\text{max}}$ for the most distant galaxies (or quasars) observable) as a function of H_0 for the analytic solutions of Einstein's cosmological equations[5] in three cases: (a) a **flat** ΛCDM (Lambda Cold Dark Matter) universe with $\Lambda > 0$; (b) an OFL (**open** Friedman–Lemaitre) universe with $k < 0$; and (c) a "**mixed**" ($\Lambda > 0$, $k < 0$) universe in a wide interval for H_0 encompassing $65 \text{ kms}^{-1}\text{Mpc}^{-1} \leq H_0 \leq 75 \text{ kms}^{-1}\text{Mpc}^{-1}$.

For a **flat universe** the analytic solutions of Einstein's cosmological equations ($\Lambda > 0$) are given by[5]

$$t(y_L) = \frac{2}{3}\left(\frac{\Lambda}{3}c^2\right)^{-1/2} y_L = \frac{2R_L}{3c}\left(\frac{R_L c^2}{2GM_L}\right)^{1/2} y_L,$$

$$R(y_L) = R_L \sinh^{2/3} y_L \qquad (11.3)$$

where $\Lambda > 0$ is Einstein's cosmological constant, c is the velocity of light, G is Newton's gravitational constant, M_L is the total mass of the universe,

$$R_L = \left(\frac{2GM_L}{\frac{1}{3}\Lambda c^2}\right)^{1/3}$$

is the cosmic radius when the cosmic density parameter Ω equals $1/2$ and y_L an auxiliary parameter $y_L \equiv \sinh^{-1}(R/R_L)^{3/2}$ going from $y_L = 0$ at the singularity (Big Bang) to $y_L \to \infty$ in the very distant future.

Then, at present,

$$R_0 = c/H_0 = R_L \sinh^{2/3} y_{L0}. \qquad (11.4)$$

It is easy to check from Eqs. (11.3) that

$$H(y_L) = \left(\frac{\Lambda}{3}c^2\right)^{1/2}\frac{\cosh y_L}{\sinh y_L},$$

$$H(y_{L0}) = \left(\frac{\Lambda}{3}c^2\right)^{1/2}\frac{\cosh y_{L0}}{\sinh y_{L0}}, \qquad (11.5)$$

$$H(y_L)t(y_L) = \frac{2}{3}\frac{y_L}{\tanh y_L},$$

$$H_0 t_0 = \frac{2}{3}\frac{y_{L0}}{\tanh y_{L0}}\text{(dimensionless)}, \tag{11.6}$$

$$\Omega(y_L) = 1 - \tanh^2 y_L,$$

$$\Omega_0 = 1 - \tanh^2 y_{L0}\text{(dimensionless)}. \tag{11.7}$$

In this model of the universe, as can be observed in Table 11.1 near the end of the chapter, the candidate most-distant galaxy (which has a redshift of 10.8) sent its light towards us well before the universe reached its Schwarzschild radius, which means that we were inside a black hole. Of course, at that time we, or rather, the seeds of the solar system if any, were also inside the same hole. At that time, for this model, the density parameter of the universe was almost

Table 11.1.   Numerical data for Fig. 11.1 (a) and (b).

$$t_0 = 13.7\,\text{Gyrs}$$
$$H_0 t_0 = 1 \longleftrightarrow H_0 = 70.856$$
$$H_0 t_0 = 1.0303 \longleftrightarrow H_0 = 73.0\ (\text{Riess } et\ al.)$$
$$H_0 t_0 = 0.9823 \longleftrightarrow H_0 = 69.6\ (\text{Bennet } et\ al.)$$
$$T_0 = 2.726\,\text{K}$$
$$R_0 T_0 = RT$$

Flat universe: $R_0 = c/H_0$, $R_{\text{Sch}} = 2GM_L/c^2$.

y_0	0.98	1.05	1.12	1.18	1.25	1.31	1.36
H_0 (km/sMpc)	62	64	66	68	70	72	74
$\Lambda \times 10^{21}$	6.86	7.9	8.91	10.0	11.0	12.1	13.2
Ω_0	0.43	0.39	0.35	0.31	0.28	0.25	0.23
T_{Sch}(K)	6.3	7	7.8	8.7	9.7	10.7	11.8
z_{Sch}	1.32	1.59	1.87	2.20	2.55	2.93	3.35

Open universe: $R_0 = (c/H_0)/\tanh y_0$, $R_{\text{Sch}} = 2GM_k/|k|c^2$.

y_0	1.59	1.81	2.06	2.41	3.03	∞
H_0 (km/sMpc)	62	64	66	68	70	71.374
Ω_0	0.15	0.10	0.062	0.032	0.009	0
T_{Sch} (K)	15.2	23.9	41.0	83.8	289	—
z_{Sch}	4.56	7.78	14.04	29.74	105	—

exactly 1, which means that the contribution of the cosmological constant (which nowadays is computed to contribute about 70%) was negligible.

Note that $\Omega = 2/3$ corresponds to the inflection point between decelerated expansion and accelerated expansion in the **flat model** (the point when the acceleration is zero). This point does not exist in the **open model**.

For an **open universe** the analytic solutions of Einstein's cosmological equations ($\Lambda = 0$, $k < 0$) are likewise given by

$$t(y_k) = \frac{R_+}{c|k|^{1/2}}(\sinh y_k \cosh y_k - y_k),$$

$$R(y_k) = R_+ \sinh^2 y_k \tag{11.8}$$

where $k < 0$ is the negative space-time curvature (which might be associated with an antigravitational radiation pressure); $R_+ = (2GM_k/|k|c^2)$ is the cosmic radius when $\Omega = 1/2$ (in this case corresponding to the cosmic Schwarzschild radius), M_k is the total mass of the universe; and y_k is the corresponding auxiliary parameter going from zero to infinity.

At present,

$$R_0 = (c/H_0)/\tanh y_0 = R_+ \sinh^2 y_{k0} \tag{11.9}$$

and then

$$H(y_k) = \frac{|k|^{1/2} c \cosh y_k}{\sinh^3 y_k},$$

$$H(y_{k0}) = \frac{|k|^{1/2} c \cosh y_{k0}}{\sinh^3 y_{k0}}, \tag{11.10}$$

$$H(y_k)t(y_k) = \frac{1}{\tanh^2 y_k} - \frac{y_k}{\tanh y_k \sinh^2 y_k},$$

$$H_0 t_0 = \frac{1}{\tanh^2 y_{k0}} - \frac{y_{k0}}{\tanh y_{k0} \sinh^2 y_{k0}}, \tag{11.11}$$

$$\Omega(y_k) = 1 - \tanh^2 y_k,$$

$$\Omega_0 = 1 - \tanh^2 y_{k0}. \tag{11.12}$$

In this model of the universe, as can be observed in Table 11.1, for $H_0 \geq 65$ kms^{-1}Mpc^{-1} the candidate most-distant galaxy (which has a redshift of 10.8) sent its light towards us after the universe reached its Schwarzschild radius (and the density parameter was equal to 1/2). At that redshift, the density parameter was smaller than 2/3 for all H_0 values analyzed, which means that the contribution of the space-time curvature was above 1/3. With this model, it may be assumed that galaxy formation started after the universe ceased behaving as a black hole.

Finally, for a **"mixed" universe** $(\Lambda > 0, k < 0)$ the interpolated solutions of Einstein's can be given by

$$\Omega_x = (1 - \tanh^2 y_L)^x (1 - \tanh^2 y_k)^{1-x}.$$

In this model of the universe, as can be observed in Table 11.1 for $x = 0.5$, the density parameter is intermediate between those of the other two models, somewhat nearer to the open than to the flat for $x = 0.5$.

We have assumed that $\Lambda = \text{const.} > 0$ for a flat universe (which leads to $\dot{R}(\infty) \to \infty$) and $k = \text{const.} < 0$ for an open universe (which leads to $\dot{R}(\infty) \to |k|^{1/2}c$, i.e. $\dot{R}(\infty) \to c$ for $k = -1$).

Figure 11.1(a) gives Ω_0 for an open (OFL) and a flat (ΛCDM) universe as a function of H_0 assuming $t_0 = 13.7$ Gyrs. H_0 as given by Riess *et al.*[2] results in $H_0 t_0$ to the right of $H_0 t_0 = 1$ while H_0 as given by Bennet *et al.*[3] results in $H_0 t_0$ to the left of $H_0 t_0 = 1$, as noted previously.

Figure 11.1(b) gives z_{Sch}, the redshift corresponding to light emitted at R_{Sch} (the radius at which the universe ceased behaving as a black hole, which could correspond to the start of galaxy formation in the open model).

The data shown in Fig. 11.1 are summarized in Table 11.1.

Note that we can approximate an average value of the density parameter $< \Omega >$ from the first galaxies to us, by computing the half sum of $\Omega(z_{\text{gal}})$ to $\Omega(0)$, with both models, for $H_0 = 69.6$. The results are $< \Omega >_{\text{flat}} = (1 + 0.2805)/2 = 0.6402$ and $< \Omega >_{\text{open}} = (0.5 + 0.0084)/2 = 0.254$, respectively.

As it is known, NASA's JWS Telescope expected to be launched on schedule in 2017 may improve considerably the accuracy with which H_0 is known. No great surprises are generally expected but certainly, reliable new data for H_0 should be welcome.

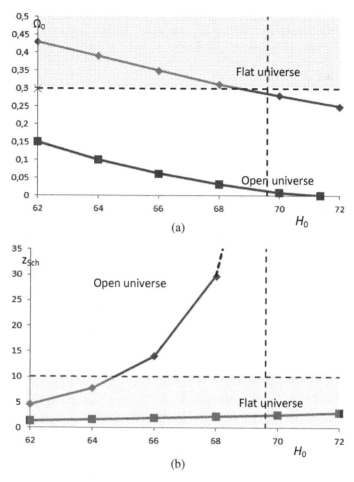

Fig. 11.1. (a) Ω_0 for an open (OFL) and a flat (ΛCDM) universe. (b) z_{Sch}, the redshift corresponding to light emitted at R_{Sch}.

References

1. L. Freedman, B. F. Madore, B. K. Gibson, L. Ferrarese, D. D. Kelson, S. Sakary, J. R. Monld, R. C. Hennicutt, H. C. Ford, J. A. Graham, J. P. Huchra, S. M. G. Hughes, G. D. Illingworth, L. M. Macri and P. B. Stetson, Final results from the Hubble Space Telescope key project to measure the Hubble constant, *Astrophys. J.* **533**, 47–72 (2001).

2. A. G. Riess, Local results, Presented in the *2014 Cosmic Distance Scale Workshop* (2014), http://www.stsci.edu/institute/conference/cosmic-distance/.
3. C. L. Bennet, D. Larson and J. L. Weiland (2014), arXiv: 1406.171v1.
4. B. M. Tinsley, The cosmological constant and cosmological change, *Phys. Today* **30**, 32–38 (1977).
5. J. A. Gonzalo and Manuel Alfonseca (2013), arXiv: 1306.0238.

Part III: More Paradoxes

My problem is solving equations but what is the meaning of the solutions... let others think about it.

Alexander Friedmann
(1888–1925)

The world is finite. Otherwise we would be obliged to affirm something which, in itself, is outside our possibility of knowledge.

Georges Lemaitre
(1894–1966)

Chapter 12

The Missing Mass and Dark Energy Paradoxes

The *critical density* is the cosmic density (mass per unit volume) at which the speed of growth of the cosmic radius ($\dot{R}$) is exactly at escape velocity, $\rho_c = 3H^2/8\pi G$, where $H = \dot{R}/R$ is Hubble's parameter. At present, the *actual cosmic density* in our close neighborhood ρ_0, estimated on the basis of ordinary matter, luminous matter from galaxies and cosmic dust, is only about 4.5% of the critical density. Many theorists, specially inflationary theory supporters, are clearly in favor of a ratio $\rho_0/\rho_c = 1$ corresponding to a flat ($k = 0$) universe. This 100% of total equivalent mass would include as much as 25% *dark matter* (of unknown provenance) and about 70% *dark energy* (of even more unknown nature) associated with the cosmological constant.

Missing mass, or dark mass, was postulated by Fritz Zwicky[1] 75 years ago to account for the non-Keplerian velocities of orbital motion around the center of mass in galaxies and galaxy clusters. Later, gravitational lensing of background cosmic objects by galaxy clusters, and the temperature distribution of hot gas within galaxies and galaxy clusters were proposed as supporting large amounts of dark mass in the universe. It should be noted that most of these reports did not consider possible cosmic density variations with distance (with increasing densities at earlier times, further distances) for sufficiently old and distant early galaxies.

After the advent of inflationary cosmology and the observational reports suggesting considerable amounts of dark matter, many cosmologists accepted the hypothesis that most matter in the universe is non-baryonic, i.e., not made up of protons and neutrons within nuclei. This mysterious dark matter was supposed not to interact with ordinary matter via electromagnetic or nuclear (strong or weak) interactions. The possibility of "hot" dark matter, "warm" dark matter, and "cold" dark matter, or a combination of the three kinds of dark matter, have been considered[2] in speculative discussions of the problem during the last four decades.

The search for tangible dark matter, however, has not been very successful.[3] According to J. Silk, cold dark matter has been seriously challenged. Some amount of non-baryonic dark matter might be homogeneously distributed throughout cosmic space, but to conclude from the observations that its amount is of the order of 26%, and that it is made up of elusive, non-tangible hypothetical particles or objects, is not very realistic. The list of potential candidates during the last three or four decades includes massive neutrinos (now more or less discarded), WIMPS (weakly interactive massive particles), MACHOS (massive compact halo objects), and massive neutron star binaries, considered to be responsible for abundant GBR (gamma-ray burst) sources in the universe. If these or other possible candidates are as abundant in the universe as to account for one quarter of its mass they should have been detected and identified long ago. It may be also pointed out that M. Milgrom[4] has proposed a modification of Newtonian dynamics as a possible alternative to the "missing" mass hypothesis. According to R. A. Alpher and R. Herman[5] it is undisputable that, at early times, the density parameter $\Omega = \rho/\rho_c$ for an open universe with $\Lambda = 0$ should be very near 1. In their words:

> We have evaluated Ω numerically and find, as expected, that at early times the value of Ω is extraordinarily close to unity and that the deviation from unity becomes increasingly small for earlier and earlier times.

As we have seen in Chapter 7, from the compact Friedmann–Lemaitre solutions [Eqs. (7.8) and (7.9)] for and open universe ($k <$ 0, or equivalently, for $\Lambda = -3k^2/R < 0$), at very early times Ω (dimensionless) approaches 1, and $H \cdot t$ (dimensionless) approaches 2/3.

Alpher and Herman conclude that:

> *It may not be necessary to invoke that consequence of an inflationary paradigm which requires that the value of Ω be unity throughout the history of the expansion ...*

Dark energy,[6] necessary to complement dark matter in order to get $\Omega = 1$, is assumed to be a certain kind of potential energy permeating all space-time and tending to increase the expansion rate at later times with respect to earlier times, the times at which the first protogalaxies were formed. It is usually assumed to be associated[7] with a nonvanishing "cosmological constant" and is sometimes connected with hypothetical scalar fields (quintessence). The cosmological constant is often tentatively explained as physically due to the vacuum energy (the zero-point energy of the vacuum). This is not so easy to do without incurring notable numerical inconsistencies in its order of magnitude. Measuring the equation of state of this dark energy is currently one of the main tasks of observational cosmology, but for the moment, the effort has not been very successful.

Thus, the dark matter and dark energy that have not yet been observed, constituting about 95% of the universe matter/energy content, which have been intensely but unsuccessfully investigated for more than twenty years, can be said to constitute one of the greatest standing cosmic paradoxes at the present time. There is the danger, however, that after so many years, cosmologists may get so used to them that begin to think that they are not at all paradoxical, but quite natural.

Would it be possible, by taking into account the *time-evolution* of the mass matter density parameter ratio $\Omega_m(y) = \rho_m(y)/\rho_c(y)$ and the space-time curvature potential energy density parameter ratio

$\Omega_k(y) = |k|(c/\dot{R}(y))^2/\rho_c(y)$, that the mysteries of the dark matter and the dark energy be somehow solved?

Equation (7.1) can be written as

$$\rho_c = \rho + \rho_c |k| \left(\frac{c}{\dot{R}(y)} \right)^2 \tag{12.1}$$

or equivalently, as

$$1 = (\Omega_m(y) + \Omega_\gamma(y)) + \Omega_k(y) = \Omega(y) + \Omega_k(y), \tag{12.2}$$

where, for $y > 0.881$ ($z > 20.7$, $T > 59.3\,\mathrm{K}$),

$$\Omega_m = \Omega \left(1 + \frac{T}{T_{\mathrm{af}}} \right) \approx \Omega = 1 - \tanh^2 y, \tag{12.3}$$

$$\Omega_k = 1 - \Omega = \tanh^2 y, \tag{12.4}$$

with $y \equiv \sinh^{-1}(T_+/T)^{1/2}$, $T_+ = 59.2\,\mathrm{K}$, $T_{\mathrm{af}} = 2968\,\mathrm{K}$.

Table 12.1.   The evolution of Ω_m (matter), Ω_r (radiation) and Ω_k (space-time curvature potential energy) as a function of time t (Gyrs) and temperature T(K).

y	(z)	t(Gyrs)	T(K)	Ω_m	Ω_k
2.243	0	13.70	2.726	0.044	0.956
2.196	0.100	12.36	3.000	0.048	0.951
2.057	0.467	9.08	4.000	0.063	0.936
1.949	0.834	7.14	5.000	0.077	0.922
1.862	1.201	5.83	6.000	0.091	0.908
1.789	1.567	4.92	7.000	0.105	0.895
1.726	1.934	4.23	8.000	0.118	0.881
1.671	2.301	3.70	9.000	0.131	0.869
1.400	4.990	1.85	16.33	0.215	0.783
1.200	8.53	1.05	26.00	0.302	0.695
1.100	11.18	0.775	33.22	0.355	0.640
0.881	**20.72**	**0.365**	**59.23**	**0.490**	**0.500**
0.800	26.54	0.266	75.09	0.545	0.440
0.700	36.75	0.173	102.92	0.613	0.365
0.600	52.60	0.106	146.20	0.678	0.288
0.400	127.80	0.030	251.23	0.765	0.144
0.140	1087.0	0.001	2968.00	0.990	0.019

Note that for $y = 0.140$, $T = T_{af} = 2968.0\,\mathrm{K}$ (decoupling, atom formation), $\Omega_m = \Omega_\gamma = 0.490$, i.e., *mass, matter* and *radiation matter* energy densities are equal and are together near one, and $\Omega_k = 0.019$ is much smaller than one.

From the beginning of the galaxy formation time ($T = T_+ = 59.2\,\mathrm{K}$) to the present ($T = T_0 = 2.726\,\mathrm{K}$), Ω_m decreases from nearly $1/2$ to nearly 0.044, and Ω_k increases from nearly $1/2$ to nearly 0.956, which lead to the following average values

$$\langle \Omega_m \rangle \approx 0.272,$$

$$\langle \Omega_k \rangle \approx 0.728.$$

The figures are surprisingly close to the current estimates for *dark matter* and *dark energy*. If the argumentation is correct (and why not) a lot of research money could be saved.

Note that, regardless of the increase in local matter mass density to be expected as $r = R_0 - R$ approaches $r = R_0 - R_+$, the actual matter mass density at any time (in a homogeneous universe) is assumed to be the same everywhere, and the total mass of the universe must therefore be the same at $t = t_+ (R = R_+)$, when the first protogalaxies begin to form, and at $t = t_o (R = R_o)$, the present: with no "missing mass", properly speaking.

The source of the space-time curvature energy density, which is $|\mathrm{k}|$, is also *independent of time*, even if Ω_k increases with time at the same rate as $\Omega = \Omega_m + \Omega_r$ decreases.

The key is to take properly into account the time-evolution of $\Omega(\mathrm{t})$ and $H(\mathrm{t})$.

In June 15, 2001 a reanalysis of WMAP's data by astronomers in the Physics Department at Durham University suggested that the conventional wisdom (dark energy 74%, dark matter 22%) about the content of the universe may be wrong.

The report[8] "Cross-correlating WMAP5 with 1.5 million LRGs: a new test for ISW effect", co-authored by astronomers from Durham University, Anglo-Australian Observatory, University of Sydney, The Pennsylvania State University and Yale University, was published in the prestigious British Journal *Monthly Notices of the Royal Astronomical Society*.

If their findings are confirmed they may end up contributing to solve satisfactorily the "dark energy" and "dark matter" paradoxes.

References

1. F. Zwicky, *Helvetica Physica Acta* **6** (1933) 110–127.
2. J. Silk, *Physics Reports* **405** (2005) 279.
3. J. Silk, *Europhysics News* **32**(6) (2001) 211.
4. M. Milgrom, *Astrophysical Journal* **270** (1983) 365–370.
5. R. A. Alpher and R. Herman, *Meeting on Unified Symmetry in the Small and the Large* (Coral Gables, Florida, 1995).
6. P. J. E. Peebles and B. Ratra, *Reviews of Modern Physics*, **75** (2003) 559–606.
7. S. Carroll, *Living Reviews in Relativity*, **4** (2001) 1.
8. U. Sawangwit *et al.*, *Monthly Notices of the Royal Astronomical Society*, **402**(4) (2009) 2228–2244.

Chapter 13

The Accelerating Universe Paradox

The paradox of the accelerated expansion of the universe can be stated as follows: if, after the Big Bang, expanding matter (mainly galaxies but perhaps all other kinds of mass matter) is moving radially away from the center of gravity of the finite mass universe, in the absence of any increasing counter-gravitational force, why are not galaxies seen as slowing down under the powerful central gravitational attraction?

The CBR, which at earlier stages provided sufficient radiation pressure to drive the expansion, is cooling down with time. So it could not be causing cosmic acceleration. That is why, when accelerated expansion was reported in 1998, the specter of the cosmological constant, which had already undergone several ups and downs in popularity, was immediately called upon to justify the observations of the Type Ia supernovae suggesting the accelerated expansion. In the following years several observations corroborated this suggestion.

Before analyzing the Type Ia supernovae observations, it is necessary to specify clearly where we take the origin of space-time. The origin of space-time should be taken at the earliest cosmic sphere lying beyond the sphere of the CBR. At the time of decoupling the primordial plasma universe became a universe of atoms and the first light (the CBR) began to propagate within the growing cosmic sphere. Our galaxy could be considered to be relatively near the origin of the spherical space occupied by the background radiation. The light emitted long ago by the supernovae in relatively far away galaxies (therefore nearer the space-time origin) suggests that the rate of change of redshift with distance was slower at $z \simeq 0.5$ than it is now at ($z \simeq 0$). What is the role to be expected for the finite speed of light in connection with this behavior? Let us see.

Information is coming out to our planet from galaxies with small redshifts (say $z = 0.01$), from farther away galaxies (with redshifts $z \simeq 0.5$), and from even more distant galaxies or quasars (up to $z \simeq 1$ or more). This reflects indirectly the time-evolution of the cosmic expansion. In order to analyze cosmic dynamics, it is necessary always to keep in mind the cosmic evolution of the main cosmic parameters, $\Omega(t) = \rho(t)/\rho_c(t)$ and $H(t) \cdot t = (\dot{R}/R) \cdot t$, where t is the time elapsed since the Big Bang, taking into account that $\dot{r}/r$ for any galaxy is not necessarily the same as $\dot{R}/R$ for the radius of the universe.

At $t < t_0$ (for instance $t = 10, 3$ Gyrs) the distance r from our galaxy to other galaxies, and the recession velocity $\dot{r}/c$ from them, would be

$$0 \leq r \leq R, \quad 0 \leq \dot{r}/c \leq 1.$$

At $t = t_0$ (i.e., now, $t_0 = 13.7$ Gyrs) the respective distance and velocity are

$$0 \leq r_0 \leq R_0, \quad 0 \leq \dot{r}_0/c \leq 1.$$

And at a time $t > t_0$ in the future (for instance $t = 15$ Gyrs) the respective distance and velocity should be

$$0 \leq r \leq R, \quad 0 \leq \dot{r}/c \leq 1.$$

At present r_0 is the distance to the supernova when the light arriving *now* to us was emitted, and the velocity $\dot{r}_0 = v_0$ (deduced from the redshift) is the recession velocity when the light arriving to us *now* was emitted. R_0 and $\dot{R}_0$ on the other hand are the cosmic radius (*now*) and its growth velocity (*now*).

Let us analyze supernovae data[8] ($1\mathrm{pc} = 3.07 \times 10^8\,\mathrm{cm}$) taking into account that $m[r(\mathrm{pc})]$, the apparent magnitude, is given by

$$m = 5\log_{10} r(\mathrm{pc}) + (M - 5) \tag{13.1}$$

where $M = M_{\mathrm{sn}} = -19.3$ is the relevant absolute magnitude to obtain the distance $r(\mathrm{pc})$. On the hand, knowing z (redshift) is sufficient to obtain

$$\frac{\dot{r}}{c} = \frac{(z+1)^2 - 1}{(z+1)^2 + 1} \le 1 \tag{13.2}$$

which gives the recession velocity divided by the speed of light. The Hubble ratio $H = \dot{R}/R$ does not necessarily coincide, as previously noted, with the ratio $h = \dot{r}/r$, except for galaxies in our close neighborhood ($z \ll 1$). For sufficiently distant supernovae, $z \approx 1$ or more, because of the relativistic relationship between $(\dot{r}/c)$ and z given above, which is a consequence of the finite value of the velocity of light, we expect that $h < H$.

Figure 13.1 represents in log-log form a set of velocity versus distance data[8] for a number of supernovae with redshifts in the interval z from 0.01 to 1. For $z \ll 1$ the velocity versus distance relationship should be given by Hubble's law $H = \dot{R}/R$ see Eq. (7.12),[1] which is specified in Fig. 13.1 as $H = 87.75$ km/s/Mpc for $t = 10.3$ Gyrs, $H_o = 67.9$ km/s/Mpc for $t_o = 13$ Gyrs (present), and $H = 61.4$ km/s/Mpc for $t = 15$ Gyrs. In all cases, we have represented $\dot{r}$ versus r taking into account the relativistic expression for $v/c \equiv \dot{r}/c$ given in Eq. (13.2).

The data points corresponding to the supernovae measured in 1998 up to $z = 1$ are well fitted by Hubble's law ($\dot{r}/r = $ constant) up to $z \approx 0.1$, but begin to deviate thereafter, confirming the expectation that (v/c) must begin to saturate towards $(v/c) \approx 1$ at some distance.

Let us see how relativistic considerations can be used to get the recession velocity in terms of $\dot{R}_o(0)$ and $\dot{R}(r_o)$, where r_o is the distance to the supernovae when the light arriving to us now was emitted. According to the Special Theory of Relativity

$$v = \frac{\dot{R}_o(0) - \dot{R}(r_o)}{1 - \dot{R}_o(0) \cdot \dot{R}(r_o)/c^2}. \tag{13.3}$$

Then, assuming that the change $d\dot{R}(r_o)$ is proportional to $-\dot{R}(r_o)dr_o$, we get

$$\dot{R}/\dot{R}_o \approx e^{-2r_o/r_m} \tag{13.4}$$

where r_m is a characteristic distance. Then

$$v \equiv v\left(\frac{r_o}{r_m}\right) = \dot{R} \cdot \left[\frac{1-\dot{R}/\dot{R}_o}{1-(\dot{R}_o^2/c^2)\cdot\dot{R}/\dot{R}_o}\right]$$

$$= \dot{R}_o \cdot \left[\frac{1+\dot{R}/\dot{R}_o}{1-(\dot{R}_o^2/c^2)\cdot\dot{R}/\dot{R}_o}\right]\left(\frac{1-\dot{R}/\dot{R}_o}{1+\dot{R}/\dot{R}_o}\right) = \alpha\tanh(r_o/r_m) \tag{13.5}$$

where, for $R \to R_o$, $(r_o \ll R_o)$ i.e., $\dot{R} \to \dot{R}_o$, we get

$$\alpha_o \approx \dot{R}_o \frac{2}{1+\left(\dot{R}_o^2/c^2\right)} \approx H_o r_m \tag{13.6}$$

and for $R \ll R_o$, $(r_o$ comparable to $R_o)$, i.e., $\dot{R} \ll \dot{R}_o$, we get

$$\alpha_m \approx \dot{R}_o \tag{13.7}$$

where $r_m = R_o - R_m$ is the supernovae distance at which the recession velocity begins to saturate towards $v = v_{\max}$. Thus, v_m must be somewhere below c, because galaxies take some time to form after the Schwarzschild radius R_+, at which $R_+ \approx c$. Then

$$\frac{v}{c} = \frac{v_m}{c}\tanh\left(\frac{r_o}{r_m}\right) \tag{13.8}$$

In Fig. 13.1 (v/c) as a function of r is given for three values of Hubble's parameter in all three cases as $(v_m/c)\tanh(\gamma r)$, where γ is the inverse of the characteristic distance r_m.

The data in Fig. 13.1 do not allow an accurate determination of the saturation velocity corresponding to the oldest observable galaxies, but they seem to indicate roughly a value $(v_m/c) \approx 0.72$ not far from one. Of course the general trend of the recession velocity of the galaxies, as shown in Fig. 13.1, implies that the rate of change of the observed galaxy velocity at distances of the order of $r \approx r_m$ is *smaller* than the rate of change of the observed galaxy velocity in our close neighborhood.

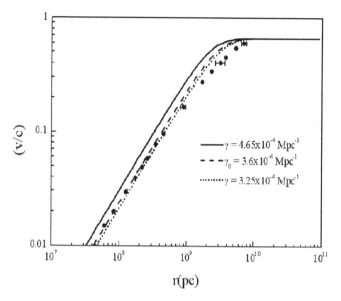

r(pc)

Fig. 13.1.   Apparent recession velocity, $\dot{r}/c = v/c$, versus apparent distance r(pc), fitted by $v/c = (v_m/c)\tanh(r/r_m) = (v_m/c)\tanh(\gamma r)$, where $\gamma \equiv 1/r_m$. Keeping only the nine points pertaining to the "High z Supernova Search" results in a much better fit with the above equation, which is not shown in Fig. 13.1 probably because α in Eq. (13.6) is somewhat larger than α in Eq. (13.7).

But it would be misleading to conclude from this that the cosmic radius $R(t)$ is really accelerating. In fact the value of the Hubble parameter, as shown in Fig. 13.1, decreases with time from $H = 87.75$ km/s/Mpc at $t = 10.3$ Gyrs, to $H_o = 67.9$ km/s/Mpc at $t_o = 13$ Gyrs (present), to $H = 61.4$ Km/s/Mpc at $t = 15$ Gyrs, as given by

$$H = \frac{\dot{R}}{R} = \frac{|k|^{1/2}c}{R_+} \cdot \frac{\cosh y}{\sinh^3 y}, \qquad (13.9)$$

obtained by means of the compact Friedmann–Lemaitre solutions of Einstein's cosmological equation in Chapter 7.

In other words, the reported acceleration of galaxies in our close neighborhood, $z \approx 0$ ($r_o \approx R_o$), is not incompatible with a really decelerating growth of the cosmic radius, as given by the Friedmann–Lemaitre solutions, Eqs. (7.10) and (7.11), and with a real decrease with time of the Hubble's parameter, as predicted by Eq. (7.12).

References

1. A. Reiss *et al.*, *Astronomical Journal* **116** (1998) 1009.
2. S. Perlmutter *et al.*, *Astrophysical Journal* **517** (1999) 565.
3. A. G. Reiss *et al.*, *Astrophysical Journal* **607** (2004) 665.
4. D. N. Spergel *et al.*, *Astrophysical Journal Suppl.* **148** (2003) 175.
5. B. Chaboyer and L. M. Krauss, *Astrophysical Journal Letters* **567** (2002) L4.
6. W. M. Wood-Vasey *et al.*, *Astrophysical Journal* **666** (2007) 694.
7. P. Astier *et al.*, *Astronomy and Astrophysics* **447** (2006) 31.
8. S. Perlmutter, *Phys. Today*, April (2003) 56.

Chapter 14

The Photon-to-Baryon
Ratio Paradox

In his book[1] *The First Three Minutes*, Steven Weinberg says:

> *We observed in the preceding chapter that there is now a nuclear particle for every 1000 million photons in the microwave radiation background (the exact figure is uncertain) so the baryon number per photon is about one billionth (10^{-9}).*

After noting that no one has seen any sign of appreciable amounts of antimatter in the universe, Weinberg considers the possibility that the density of photons has not remained proportional to the inverse cube of the size of the universe:

> *This could happen if there were some sort of departure from thermal equilibrium, some sort of friction or viscosity which could have heated the universe and produced extra photons. In this case, the baryon number per photon might have started at some reasonable number, perhaps around one, and then dropped to its present low value as more photons were produced.*

This could be called the *baryon-to-photon paradox* (10^{-9}) or the *photon-to-baryon paradox* (10^9).

At the time this was written the expansion and cooling down of the universe was generally assumed to be described by a *single equation of state*. However, this was not very realistic.[2]

There is a distinct change in the universe at T_{af} (atom formation), i.e., about 3 000 to 4 000 degrees K, from a *plasma universe* $(T > T_{af})$ to an *atom universe* $(T < T_{af})$. The former is *opaque* to radiation, therefore, in it charged nuclei (mainly H and ^{4}He nuclei) and electrons scatter photons violently (friction) and extra photons are continuously produced. On the other hand, the atomic universe (at which, in due time, stars and galaxies would form) is basically *transparent*. Then the conservation of the number of photons (as well as that of stable nuclei) is to be expected thereafter through the expansion.

This implies a change in *equation of state* from the plasma universe, in which radiation mass density and matter mass density expand in *unison* $(T > T_{af})$ as given by

$$\rho_r(R) = \rho_m(R), \text{ resulting in } RT^{4/3} = R_{af}\, T_{af}^{4/3} \qquad (14.1)$$

of the present "atomic" universe, in which the number of photons $n_r(R)$ and of baryons $n_b(R)$ are both conserved through the expansion $(T < T_{af})$ and are given by

$$n_r(R) = n_b(R), \text{ resulting in } RT = R_{af}\, T_{af}. \qquad (14.2)$$

This implies that the number of photons equals the number of baryons at the characteristic baryon temperature $T_b = m_b c^2/2.8\, k_B$, $T > T_{af}$ (plasma universe) and we have:

$$\frac{n_r}{n_b} = \frac{\sigma T^4/2.8\, k_B T}{\left[M/\frac{4\pi}{3}R^3\right]/m_b} = \left(\frac{\sigma T^4}{\left[M/\frac{4\pi}{3}R^3\right]\cdot c^2}\right)\frac{m_b c^2/2.8\, k_B}{T} = \frac{T_b}{T} \qquad (14.3)$$

since $\rho_r = \rho_m$ during the plasma phase of the universe $(T > T_{af})$. Therefore, at $T \approx T_b$ (baryon equilibrium temperature) we have

$$\frac{n_r(T_b)}{n_b(T_b)} \approx 1 \qquad (14.4)$$

as might be expected.

The cosmic *neutron-to-proton* ratio determined from the *relative abundance* of ^{4}He in the universe is

$$(n/p)_{\text{obs}} \approx 0.13 \pm 0.02. \tag{14.5}$$

If we assume that it is due to the spontaneous disintegration of primordial neutrons ($T_n = m_n c^2/2.8\, k_B \approx 3.88 \times 10^{12}\,$K) into protons, electrons and neutrinos once cosmic conditions are apt to allow free electrons ($T_e = m_e c^2/2.8\, k_B \approx 2.11 \times 10^9\,$K) we can use the known half-life of the neutron $\tau = 887\,$s to calculate the neutron-to-proton ratio by means of

$$(n/p) \approx e^{-(t_{\text{ns}} - t_e)/\tau} \tag{14.6}$$

where t_{ns} is the cosmic time at the end of nucleosynthesis ($T_{\text{ns}} \approx 0.5 \times 10^9\,$K), with an estimated error of $\pm 20\%$, t_e is the cosmic time at the electron equilibrium temperature, and τ the free neutron half-life.

At this early epoch ($y \ll 1$) cosmic time is given by

$$t(y) \approx \frac{R_+}{|k|^{1/2}c} y^3 \tag{14.7}$$

which, taking into account that

$$y \equiv \sinh^{-1}\left[\frac{R}{R_+}\right]^{1/2} = \sinh^{-1}\left[\left(\frac{T_+}{T_{\text{af}}}\right)\left(\frac{T_{\text{af}}}{T}\right)^{4/3}\right]^{1/2}$$

$$\approx \left(\frac{T_+}{T_{\text{af}}}\right)^{1/2}\left(\frac{T_{\text{af}}}{T}\right)^{2/3} \tag{14.8}$$

reduces to

$$t = \left\{\frac{R_+}{|k|^{1/2}c}\left(\frac{T_+}{T_{\text{af}}}\right)^{3/2}\right\}\left(\frac{T_{\text{af}}}{T}\right)^2 = 6.09 \times 10^{13}\left(\frac{2968}{T}\right)^2 \tag{14.9}$$

where $R_+ = 4.588 \times 10^{26}\,$cm, $|k^{1/2}| = 0.7071$, $c = 3 \times 10^{10}\,$cm/s, $T_+ = 59.2\,$K and $T_{\text{af}} = 2968.$[3]

The cosmic times at neutron formation (t_n), electron formation (t_e) and at the end of nucleosynthesis (t_{ns}) would then be given by

$$t_n = 3.56 \times 10^{-5}\,\text{s} \tag{14.10}$$

$$t_e = 120.5\,\text{s} \tag{14.11}$$

$$t_{ns} = 2145 \pm 400\,\text{s}. \tag{14.12}$$

Therefore

$$(n/p)_{\text{cal}} \approx e^{-(2145 \pm 400 - 120.5)/887} = 0.10 \pm 0.04, \tag{14.13}$$

which agrees reasonably well with the observed ratio given in Eq. (14.5).

We can check also that with Eq. (14.1), the cosmic equation of state for a *plasma universe*, the ratio of cosmic radii for closely packed electrons at T_e and closely packed baryons at T_b

$$\frac{R_e}{R_b} = \left(\frac{T_b}{T_e}\right)^{4/3} = \left(\frac{m_b}{m_e}\right)^{4/3} \approx 2 \times 10^4, \tag{14.14}$$

is approximately the same as the ratio of particle radii. (For the electron radius taking into account Pauli's principle, we estimate $r_e \approx (1/2) \cdot r_{\text{Bohr}} \approx (1/2) \cdot 0.53 \times 10^{-8}$ cm. For the baryon radius we take $r_b \approx 1.24 \times 10^{-13}$).[4]

$$\frac{r_e}{r_b} \approx \frac{0.26 \times 10^{-8}}{1.24 \times 10^{-13}} \approx 2 \times 10^4, \tag{14.15}$$

which is not the case using Eq. (14.2), the cosmic equation of state for a *transparent universe*, which would result in

$$\frac{R_e}{R_b} = \left(\frac{T_b}{T_e}\right) = \left(\frac{m_b}{m_e}\right) = 1836 \ll \frac{r_e}{r_b} \approx 2 \times 10^4. \tag{14.16}$$

During the transparent phase of the expansion, radiation density (ρ_r) decreases much faster than matter density (ρ_m). In fact, at present $\rho_{mo}/\rho_{ro} \approx (T_{af}/T_0) \approx 1088$ within the same spherical spacetime. Where is the missing radiation energy going? Let us examine what happens with the cosmic zero point energy.

References

1. S. Weinberg, *The First Three Minutes* (Bantam Books, New York, 1977), pp. 88–89.
2. N. Cereceda *et al.*, Is it realistic assume the same cosmic equation of state prior to and after atom formation? Frontiers of Fundamental Physics, Universidad Politécnica, Madrid 2006, *AIP Conference Proceedings*, **905**.
3. WMAP Collaboration, *Astrophys. J. Suppl.* **148** (2003) 13–15, 17, 18, 20, 22–25.
4. R. D. Evans, *The Atomic Nucleus* (McGraw-Hill, New York, 1955).

Chapter 15

Cosmic Zero-Point Energy

The **zero point energy**[1] for electromagnetic radiation confined in a finite expanding spherical cavity of radius R, no matter how large, is given by

$$\Delta \bar{E}_0(R) = \sum_{\bar{k}} \frac{1}{2} h v_{\bar{k}} \cong \int_{v_{\min}(R)}^{v_{\max}(R_{\min})} V \cdot \frac{h}{c^3} v_{\bar{k}}^3 \mathrm{d}v_{\bar{k}} \qquad (15.1)$$

where $V \cong L^3 \cong \frac{1}{2}(2R)^3$, and $v_k = \frac{c}{R}$, $\mathrm{d}v_k = \frac{c}{R^2}\mathrm{d}R$.

Therefore taking $R_{\min} = R_H = h/Mc$, which is Heisenberg's radius (or Compton's radius) for a universe of mass M, enormous but finite, we have (except for a factor of order unity)

$$\Delta \bar{E}_0(R) > -ch \int_R^{R_H} \frac{\mathrm{d}R}{R^2} = -ch \left[-\frac{1}{R} \right]_R^{R_H} = ch \left[\frac{1}{R_H} - \frac{1}{R} \right]; \quad (15.2)$$

and taking into account that $ch/R_H = Mc^2$, for $R \gg R_H$ the zero point energy would be given by

$$\Delta \bar{E}_0(R) \cong \frac{ch}{R_H} \left[1 - \frac{R_H}{R} \right] \cong Mc^2. \quad (R \gg R_H) \qquad (15.3)$$

In order to see how $\bar{E}_0(R)$ evolves with time from $t = t_H = 0.46 \times 10^{-102}$ s, to "decoupling" (atom formation), $t_{dec} \cong 10^{14}$ s, i.e. during the **plasma phase** of cosmic expansion at which ρ_r (radiation) $= \rho_m$ (matter), as discussed below and then from $t_{dec} \cong 10^{13}$ s to the present epoch $t_0 = 13.7 \times 10^9$ yrs, i.e. during the **transparent phase** of the expansion, at which ρ_r decreases continuously from $\rho_{rdec} = \rho_{mdec}$ to its present value $\rho_{ro} \ll \rho_{mo}$, it is necessary to specify the **equation of state**,[2,3] $R(T)$, for the **plasma phase** and for the **transparent phase** in which, at some time stars and galaxies begin to form.

In the **plasma phase**, after massive charged particles and electrons are formed, radiation and particles move in unison away from the center of the sphere specified by t_H (almost indistinguishable from the vanishing sphere specified by $t = 0$, the Big Bang singularity). In due time material particles — protons, ^{4}He nuclei, electrons scatter photons (radiation) and photons push away charged particles resulting in cosmic expansion.

Before decoupling (atom formation) and at least since primordial nucleosynthesis (^{4}He formation), i.e. since $T_{ns} \cong 4.6 \times 10^8$ K,

$$\rho_r(T) = \sigma T^4 = \rho_m = Mc^2 \left/ \frac{4\pi}{3} R^3 \right., \tag{15.4}$$

which gives the **plasma phase equation of state**

$$R^3 T^4 = Mc^2 \left/ \frac{4\pi}{3} \sigma \right. = \text{const.} \tag{15.5}$$

leading to

$$\Delta \bar{E}_0 = \bar{E}_0(R) - \bar{E}_0(R_H) = Mc^2 \left[1 - \frac{R_H}{R} \right]$$

$$= Mc^2 \left[1 - \left(\frac{T}{T_H} \right)^{4/3} \right] \cong Mc^2. \tag{15.6}$$

On the other hand, in the *transparent phase*, after decoupling (atom formation), $T_{af} \cong 3000$ K, neutral atoms have been formed, violent scattering of photons has practically ceased, and the total number of photons (as the total number of baryons tied up in atoms, mainly H and ^{4}He nuclei at the beginning) becomes fixed in the universe. Then,

$$n_\gamma(T) = \frac{\sigma T^4}{2.8 k_B T}, \quad n_m = \frac{Mc^2 \left/ \frac{4\pi}{3} R^3\right.}{m_b},$$

$$(n_\gamma/n_m = \text{const.}, \quad t > t_{dec}) \tag{15.7}$$

which results in the **cosmic equation of state at the present phase**

$$R^3 T^3 = \text{const.} \tag{15.8}$$

resulting in $\rho_\gamma = (\rho_\gamma)_{dec}(T/T_{dec})$, and, therefore

$$\Delta \bar{E}_0 = \bar{E}_0(R) - \bar{E}_0(R_{dec}) = Mc^2 \left[1 - \frac{T}{T_{dec}}\right] < Mc^2. \tag{15.9}$$

Equation (15.8) means that from T_{dec} to T_0 the radiation energy density decreases with respect to the matter energy density as $\rho_r/\rho_m \cong T/T_{dec}$.

After decoupling (atom formation)

$$\rho_\gamma = \rho_r(T_{dec}) \left(\frac{T}{T_{dec}}\right)^4,$$

$$\rho_m = \rho_m(R_{dec}) \left(\frac{R_{dec}}{R}\right)^3,$$

and $RT = R_{dec} T_{dec} = \text{const.}$,

$$\frac{\rho_r}{\rho_m} = \left(\frac{\rho_r}{\rho_m}\right)_{dec} \left(\frac{T}{T_{dec}}\right) = \frac{T}{T_{dec}}. \tag{15.10}$$

Denoting by ρ_{zp} the "zero point" energy density

$$\rho_{zp} = \rho_r(T_{dec}) \left(1 - \frac{T}{T_{dec}}\right),$$

$$\rho_m = \rho_m(R_{dec}) \left(\frac{R_{dec}}{R}\right)^3,$$

and again $RT = R_{dec} T_{dec} = \text{const.}$,

$$\frac{\rho_{zp}}{\rho_m} = \left(\frac{\rho_r}{\rho_m}\right)_{dec} \left(1 - \frac{T}{T_{dec}}\right) = 1 - \frac{T}{T_{dec}}. \tag{15.11}$$

Putting together Eqs. (15.10) and (15.11) we get

$$\frac{\rho_r}{\rho_m} + \frac{\rho_{zp}}{\rho_m} = \left[\frac{T}{T_{dec}} + \left(1 - \frac{T}{T_{dec}} \right) \right] = 1$$

which ensures energy conservation *taking into account the temperature dependence of the finite cosmic zero-point energy* throughout the expansion. In other words,

$$\Delta\rho_r + \Delta\rho_{zp} = \Delta\rho_m, \quad \text{for all } T. \tag{15.12}$$

References

1. R. Loudon, *The Quantum Theory of Light*, 2nd Ed. (Clarendon Press, Oxford, 1983), p. 139.
2. N. Cereceda, M. I. Marqués, G. Lifante and J. A. Gonzalo, in *Frontiers of Fundamental Physics: Eighth International Symposium FFP8*, Madrid, Spain, 17–19 October 2006, B. G. Sidharth, A. A. Faus and M. J. Fullana (eds.) (American Institute of Physics, 2007), AIP Conf. Proc., Vol. 905, pp. 6–12.
3. Julio A. Gonzalo, *The Intelligible Universe*, 2nd Ed. (World Scientific, Singapore, 2008), pp. 312–317.

Part IV: A Contingent Universe

In a world that was not the expression of intelligence science could never have come into being.

Edmund Whittacker
(1873–1956)

Chapter 16

The Universe is Finite, Open and Contingent

The universe is *finite*: As we have seen, it has a finite *mass*

$$M = 1.54 \times 10^{54} \, \text{g}, \tag{16.1}$$

a finite (but growing) *age* since the Big Bang,

$$t_o = 13.7 \times 10^9 \, \text{yrs}, \tag{16.2}$$

and a finite (but growing) *radius*, since that event,

$$R_o = 9.96 \times 10^{27} \, \text{cm}. \tag{16.3}$$

The finite mass $M = 1.54 \times 10^{54}$ g is conserved, of course through the cosmic expansion, while the actual time and the actual radius are finite, but increasing and, in an *open* universe, unbounded. We may note that the characteristic radius R_+, for a universe with a total matter mass given by $M = 1.54 \times 10^{54}$ g, with $y_+ = \sinh^{-1}$,[1] is

$$t(y_+) = \frac{R_+}{|k|^{1/2}c}\{\sinh y_+ \cosh y_+ - y_+\} = 0.365 \times 10^9 \, \text{yrs}, \tag{16.4}$$

$$R(y_+) = R_+ \sinh^2 y_+ = 4.58 \times 10^{26} \, \text{cm}. \tag{16.5}$$

A *Compton* radius

$$r_c = \frac{\hbar}{mc}, \tag{16.6}$$

and a *Schwarzschild* radius

$$r_s = \frac{Gm}{c^2}, \tag{16.7}$$

can be associated to the universe as well as to various finite objects in it, going from finite galaxies to finite elementary particles.

Table 16.1 gives *Compton* radius and *Schwarzschild* radius for various massive objects in the universe.

As we saw in Chapter 7, Eq. (7.16), if the Friedmann–Lemaitre solutions of Einstein's cosmological equations describe correctly cosmic evolution (and they do describe well the thermal history of the universe from the Big Bang to the present), the dimensionless product of the Hubble constant time the present age of the universe is given by

$$H_o t_o = 0.942 \pm 0.065, \tag{16.8}$$

which is incompatible either with a flat universe ($k = 0$), which requires

$$H_o t_o = 2/3, \tag{16.9}$$

or with a closed universe ($k > 0$), which requires

$$H_o t_o < 2/3. \tag{16.10}$$

The universe is therefore *open*, finite and unbounded. Modern physics tells us that we live in an evolving, finite and open universe.

Why is the universe what it is according to modern cosmology, and not anything else?

Table 16.1.   Compton and Schwarzschild radii for massive objects.

Object	m(g)	r_c(cm)	r_s(cm)	r_c/r_s
Universe	$1.54 \cdot 10^{54}$	$2.26 \cdot 10^{-92}$	$1.14 \cdot 10^{26}$	$1.98 \cdot 10^{-118}$
Galaxy	$\sim 1 \cdot 10^{43}$	$3.5 \cdot 10^{-81}$	$7.41 \cdot 10^{-14}$	$0.47 \cdot 10^{-95}$
Star	$\sim 1 \cdot 10^{32}$	$3.5 \cdot 10^{-70}$	$7.41 \cdot 10^{3}$	$0.47 \cdot 10^{-73}$
Earth	$5.95 \cdot 10^{24}$	$5.85 \cdot 10^{-63}$	$4.43 \cdot 10^{-4}$	$1.32 \cdot 10^{-67}$
Planck monopole	$2.17 \cdot 10^{-5}$	$1.61 \cdot 10^{-33}$	$1.61 \cdot 10^{-33}$	1
Baryon	$1.67 \cdot 10^{-24}$	$2.09 \cdot 10^{-14}$	$1.23 \cdot 10^{-52}$	$1.23 \cdot 10^{38}$
Electron	$9.10 \cdot 10^{-28}$	$3.84 \cdot 10^{-11}$	$6.74 \cdot 10^{-56}$	$0.56 \cdot 10^{45}$

No purely physical theory and no purely physical experiment can give a concrete answer.

In other words the universe is *contingent*, it is not *necessarily* what it is, but it is *really* what it is, and not anything else. The term "contingent" implies a *physical* reality which cannot be measured directly, and also a *metaphysical* reality, more intangible but no less real, and which can be intellectually recognized.

And a contingent universe is a created universe.

By the middle of the nineteenth century,[1] both the Hegelian *left* (Marx and Engels) and the Hegelian *right* (specially the neo-Kantians) had for a fundamental tenet that the universe (material or not) was *infinite*. Only a few first-rate scientists dared to disagree. Among them is *Gauss*, the prince of mathematicians, who noted that *Kant*'s dicta on categories were sheer triviality, probably keeping in mind non-Euclidean geometries. The finiteness of matter in endless space was implicit also in the work of *Riemann* and Zöllner.

Kant's *claim*[1] that the universe was a bastard product of the metaphysical cravings of the human intellect (put forward to discredit the classic cosmological argument to prove God's existence) was flatly denied with words and deeds by Planck and Einstein, the two greatest physicists of the twentieth century.

Planck, after liberating himself of Mach's tutelage, when he affirmed unambiguously[2] his full confidence in the reality of a *causally connected universe*.

Einstein, no less after finally emancipating himself from Mach's influence, produced the first contradiction-free treatment[1] of the totality of all gravitationally interacting objects which explicitly required a *universe with a finite mass*.

When Planck's son, Erwin, was executed for plotting against Hitler at the end of World War Two, everything seemed to have fallen in ruins around him: home, country, science. He wrote to a friend[3]:

What helps me is that I consider it a favor of heaven that since childhood a faith is planted deep in my innermost being, a faith in the Almighty and the All-good not to be shattered by anything. Of course his ways are not our ways, but trust in him helps us through the darkest trials.

In 1952, few years before his death, Einstein wrote to his friend Maurice Solovine[4]:

> ...that I think of the comprehensibility of the world... as a miracle (emphasis added) or an eternal mystery. But surely, a priori, one should expect the world to be chaotic. One might... expect that the world evidenced itself as lawful only so far as we grasp it in an orderly fashion. On the other hand, the kind of order created, for example, by Newton's gravitational theory is of a very different character... Therein lies the "miracle" which becomes more and more evident as our knowledge develops... And here is the weak point of positivists and professional atheists, who feel happy because they think that they have preempted not only the world of the divine but also of the miraculous...

As Stanley L. Jaki, perhaps the most incisive historian of Science of the twentieth century, writes[5] in his book *The Roads of Science and the Ways to God*, Planck and Einstein, with their confidence in the reality of a causally connected and finite universe, provide extraordinary compelling evidence in favor of a realistic metaphysics and epistemology, midway between idealism and positivism.

References

1. S. L. Jaki, *God and the Cosmologist* (Real View Books, Fraser, Michigan, 1998), pp. 13, 14, 17.
2. M. Planck, *A Survey of Physics*, translated by R. Jones and D. H. Williams (Methuen, London, 1925), pp. 1–41.
3. A. Herman, *Max Planck in Selbstzeungnissen und Bilddokument*, p. 98.
4. A. Einstein, *Lettres a Maurice Solovine* (Gauthier Villars, Paris, 1956), p. 102.
5. S. L. Jaki, *The Road of Science and the Ways to God* (University of Chicago Press, 1979), pp. 165–196.

Chapter 17

The Very Early Universe: Indeterminacy or Uncertainty

Quantum Mechanics, invented by de Broglie, Schrödinger, Heisenberg, Born, among others, to describe the physics of atoms and nuclei, is a statistical branch of physics,[1] and, as such, not very apt to describe the physics of the universe, a *singular entity*, absolutely unique, in spite of the current efforts to disguise it in a "multitude", called multiverse.

Heisenberg's principle,[2] which in 1927 specified correctly the inherent limitations to the accuracy of *complementary measurements* in the microscopic world of atoms and elementary particles (for instance position and momentum, time and energy), is based on principle's validity upon the specific value of Planck's constant, h or $\hbar = 1.05 \times 10^{-27}$ erg sec (the indivisible "quantum" of action). No doubt Heisenberg's principle has been extremely useful in describing the physical behavior of atoms, nuclei and particles; it can provide also unexpected and very significant insights about macroscopic physical systems; but has been also much abused by physicists, including Bohr and Heisenberg himself. Already in the very paper introducing his principle Heisenberg concluded that "the invalidity or at least the meaninglessness of the law of causality seems to be firmly established through recent developments in atomic physics". He said, further, that the inability of physicists to measure nature *exactly* implied the inability of events to take place in nature *exactly*. As noted by Jaki,[2] here the same word "exactly" is taken in two very different meanings, one operational (observational) and one ontological (factual, independent of the observer). The first does not exclude the second. As it is well known, Einstein, and with him many other great twentieth century physicists, including de Broglie and Schrödinger, did not agree with Heisenberg in the arbitrary invalidation of the law of *causality* in nature. (Where could we, poor physicists, go in the investigation of nature without a reasonable *confidence* — Planck's word — in the reality of a *causally connected universe?*).

Very conspicuous cosmologists, like Stephen Hawking, and many others, have tried by all means to suppress a Creator and an origin in time for the universe. One of the tricks is using surreptitiously Heisenberg's principle to postulate a virtual fluctuation of energy for an infinitesimal period of time, and jump from there to an actual physical universe existentially stable for billions of years.

Sometimes Heisenberg's principle is called correctly as the *uncertainty principle*. After all, the only means we have at our disposal to measure electrons and photons are photons and electrons. Limitations in the ultimate accuracy of the measurements are to be expected. Sometimes, however, Heisenberg's principle is called, improperly, the *indeterminacy* principle, implying that physical reality is undecided, not knowing for sure where *to be* or where *to go* in space-time at any moment. This is not serious. Applying this way of reasoning to a microscopic event, such as the disintegration of an alpha particle, could be taken as a sleight-of-hand or a joke. But applying it to the universe is really too much.

Bohr, who was Heisenberg's mentor, did not see any problem when he said "one speaks of a choice on the part of nature".[3] No wonder that he choose the *yin* and *yang* as his coat of arms, demonstrating that the Copenhagen interpretation of Quantum Mechanics (philosophy, not physics) implied a shift from Western to Eastern cultural references. He held also that a great truth is a statement whose contradiction is also a great truth. This sounds enlightened but in reality it is not.

Regarding the singularity implied by Einstein's cosmological equation with respect to *time*, as it was noted in Chapter 4, Heisenberg's uncertainty principle provides a limit to how close one is allowed to approach the origin. It is

$$t_H \equiv \frac{\hbar}{M_u c^2} \approx 7.54 \times 10^{-103} \, \text{s},$$

determined by the finite mass of the universe $M_u = 1.54 \times 10^{54}$ g and the values of the universal constants $\hbar$ and c. This is very small in comparison with Planck's time

$$t_{\text{Pl}} \equiv (\hbar G/c^5)^{1/2} = 5.37 \times 10^{-44} \, \text{s}.$$

No cosmic clock will ever be available to measure time at t_h, but it does help to know that there is an earliest limit.

References

1. G. 't Hooft, Plenary lecture at the International Symposium of "Frontiers of Fundamental Physics", Universidad Politecnica, Madrid, 2006, *AIP Conf. Proc.* **905** (2007), pp. 84–102.
2. S. L. Jaki, *God and the Cosmologists* (Real View Books, Fraser, Michigan, 1998).
3. S. L. Jaki, *The Ways to God and the Road of Science* (Chicago University Press, Chicago, 1987), p. 203.

Chapter 18

Why an Open ($k{<}0$) Cosmic Model is Better

Introduction

When cosmic inflation in a flat ($k = 0, \Lambda > 0$) universe was postulated by Alan Guth, about thirty years ago, the time elapsed between the Big Bang and today (t_0) was estimated as being between 10 and 20 gigayears, and the present Hubble's ratio ($H_0 = \dot{R}_0/R_0$) was considered to be between 50 and 100 km/s/Mpc. Data from the Hubble, COBE, WMAP and Planck satellites reduced the uncertainties to less than one percent in t_0 and a few percent in H_0. It is shown here that results from an open, energy-conserving cosmic model are self-consistent and compatible with the latest measurements $t_0 = 13.8 \times 10^9$ yr (WMAP, Planck) and $H_0 = 67.1$ km/s/Mpc (Bennet et al. estimate the error to be ± 2.8 km/s/Mpc); while results from a flat, inflationary model are not so. Cosmic acceleration as reported at galactic redshifts $z \approx 1$, galaxy formation times, cosmic growth from very early time ($t_{\text{HL}} \ll t_{\text{Planck}}$) to $t_{\text{infl}} \approx 10^{-35}$ s, data at "last scattering" (decoupling), and cosmic zero-point energy temperature dependence after decoupling are shown to fit better with an open rather than a flat cosmic model.

Open ($k < 0$) Cosmic Model

Let us first investigate the dependence on y_0 of the dimensionless product $H(y_0)t(y_0) = H_0 t_0$ as given[1] by

$$H(y_0)t(y_0) = \frac{1}{\tanh^2 y_0} - \frac{y_0}{\tanh y_0 \sinh^2 y_0} \qquad (18.1)$$

for an open ($k < 0$) cosmic model. As mentioned in the Introduction we assume $t_0 = 13.8 \times 10^9$ yr $= 4.351 \times 10^{17}$ s and $H_0 = 67.1$ km/s/Mpc $= 2.176 \times 10^{-18}$ s^{-1}. The accompanying Table 18.1 gives $H_0 t_0$ and $\Omega_0 = 1/\cosh^2 y_0$ as a function of y_0 in the interval of H_0 between 65.0 and 69.5 km/s/Mpc.

The observed ratio $\Omega_{m_0} = \Omega_0 = \rho_0/\rho_{c_0}$ (actual mass density to critical (escape) mass density) in our local neighborhood coincides with the purely baryonic density $(\Omega_{m_0})_b \sim 0.044$ from which we get $y_0 = \cosh^{-1}(1/\Omega_{m_0}) = 2.241$, which implies $H_0 = 2.176 \times 10^{-18}s^{-1} = 67.1$ km/s/Mpc, within 3.5 percent of Bennet *et al.* (2014)'s value. This corresponds to $H_0 t_0 = 0.940$.

We can check the internal consistency using the open ($k > 0$, $\Lambda = 0$) solutions of Einstein's cosmological equation for a finite universe[a] as follows:

$$R_0 = R_+ \sinh^2 y_0 = c/H_0 = (3 \times 10^{10})/(2.176 \times 10^{-18})$$
$$= 1.378 \times 10^{28} \text{ cm} \qquad (18.2)$$

Table 18.1. $H_0 t_0$ and $\Omega_0 = 1/\cosh^2 y_0$ as a function of y_0 in the interval of H_0 between 65.0 and 69.5 km/s/Mpc.

y_0	2	2.2	2.4	2.6	2.8	3.0
$H(y_0)t(y_0)$	0.9182	0.9368	0.9516	0.9636	0.9728	0.9798
Ω_0	0.0706	0.0479	0.0328	0.0218	0.0146	0.0098
H_0	65.05	66.37	67.42	68.27	68.92	69.42

---

[a]Note that we can estimate very accurately $H_0 = \dot{R}_0/R_0$ for $y \approx y_0$ because $H_0 = \dot{R}_0(1+\delta\dot{R}_0/\dot{R}_0)/R_0(1+\delta R_0/R_0)$ where $\delta\dot{R}_0/\dot{R}_0 \ll 1$ and $\delta R_0/R_0 \ll 1$, and therefore $R_0 \approx c/H_0$. This must be expected to become less accurate for $z \gg 1$, when relativistic effects become increasingly important.

where $R_+ \equiv R_{Sch}$, and therefore

$$R_+ = (2GM_u/|k|c^2) = R_0/\sinh^2 y_0 = 6.381 \times 10^{26} \text{ cm}. \qquad (18.3)$$

Using $G = 6.67 \times 10^{-8}$ cgs, $c = 3 \times 10^{10}$ cm/s, we can get the total mass of the universe for $k < 0, \Lambda > 0$ as shown below.

The present cosmic time is given by

$$t_0 = \frac{R_+}{c|k|^{1/2}} [\sinh y_0 \cosh y_0 - y_0] = 13.8 \times 10^9 \text{ yr} = 4.351 \times 10^{17} \text{ s}$$

$$(18.4)$$

and, using $y_0 = 2.241$, i.e. $[\sinh y_0 \cosh y_0 - y_0] = 19.89$,

$$R_+ = (4.3519 \times 10^{17}) c|k|^{1/2} \qquad (18.5)$$

which, together with Eq. (18.3), leads to

$$|k|^{1/2} = 0.975 \to |k| = 0.950 \qquad (18.6)$$

which implies $|k| \approx 1$, $k \approx -1$.

Finally, Eqs. (18.3) and (18.6) result in

$$M_u = \frac{R_+|k|c^2}{2G} = 4.10 \times 10^{54} \text{ g} \qquad (18.7)$$

for an open universe.

Flat $(k = 0, \Lambda > 0)$ Cosmic Model

Next we investigate the case for a flat cosmic model assuming $t_0 = 13.8 \times 10^9$ yr, $H_0 = 67.1$ km/s/Mpc, $H_0 t_0 = 0.940$. As it will be noted later the results are not very sensitive to the given H_0 value.

Recall that in this flat universe $H_0 t_0$ is given by

$$H_0 t_0 = \frac{2}{3} \frac{y_0}{\tanh y_0} = 0.940 \qquad (18.8)$$

which implies

$$y_0(\text{flat}) = 1.16 \qquad (18.9)$$

and

$$\Omega_{m_0}(\text{flat}) = \frac{1}{\cosh^2 y_0} = 0.321. \qquad (18.10)$$

In a flat universe the fraction of dark matter (missing, non-baryonic) matter) is very large: $\Omega_{m_0}(\text{non-baryonic}) = 0.312 - 0.044 = 0.277$.

The present cosmic radius is given by

$$R(y_0) = R(\Lambda) \sinh^{2/3} y_0 = c/H_0 = 1.378 \times 10^{28}\,\text{cm} \qquad (18.11)$$

and therefore

$$R(\Lambda) = R_0/\sinh^{2/3} y_0 = 1.081 \times 10^{28}\,\text{cm}. \qquad (18.12)$$

We can calculate the total mass of the universe in a flat model as follows.

According to Ref. 1,

$$t_0(y_0) = \frac{2}{3}\left(\frac{1}{\frac{\Lambda}{3}c^2}\right)^{1/2} y_0 = \frac{2}{3}\left(\frac{R(\Lambda)}{2GM_u}\right) y_0, \qquad (18.13)$$

and we have also that

$$t_0 = \frac{2}{3}\left(\frac{R^3(\Lambda)}{2GM_u}\right)^{1/2} y_0 = 13.8 \times 10^9\,\text{yr} = 4.351 \times 10^{17}\,\text{s}. \qquad (18.14)$$

Therefore, in this flat universe we have

$$M_u = \left(\frac{R(\Lambda)}{[\frac{3}{2}\frac{t_0}{y_0}(2G)^{1/2}]^{2/3}}\right)^3 = 2.869 \times 10^{55}\,\text{g}, \qquad (18.15)$$

comparable, but appreciably larger than that for an open universe.

Discussion

It is well known that $r = R_0 - R$ and $\dot{r} = \dot{R} = v$ are related to the redshift (z) of a receding galaxy by means of

$$r = R_0 \frac{z}{1+z}\,(r \approx R_0 z \text{ for } z \ll 1), \qquad (18.16)$$

$$v = c\frac{(1+z)^2 - 1}{(1+z)^2 + 1}\,(v \approx cz \text{ for } z \ll 1). \qquad (18.17)$$

After decoupling and atom formation, galaxies can be expected to begin forming at a time[b] t_{Sch} (Schwarzschild) when the expanding cosmic sphere is no longer a "black hole", i.e. $R \geq R_{\text{Sch}} = 2GM_u/c^2$. No galactic redshifts $z > z_{\text{Sch}}$ can be observable, for a flat as well as for an open cosmos.

For an open universe ($y_0 = 2.241$),

$$z(y) = [1/(\sinh y/\sinh y_0)^2] - 1 \leq z(y_{Sch} = 0.881) = 20.62. \quad (18.18)$$

For a flat universe ($y_0 = 1.16$),

$$z(y) = [1/(\sinh y/\sinh y_0)^{2/3}] - 1 \leq z(y_{Sch} = 0.244) = 2.24. \quad (18.19)$$

Figure 18.1 gives $\log_{10}(r/R_0)$ vs $\log_{10}(v/c)$ for an open universe around $z \approx 1$, showing an upwards trend in $\log_{10}(r/R_0)$, which is proportional to the apparent magnitude (m). This upwards trend[c] is

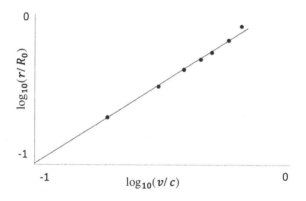

Fig. 18.1. $\log_{10}(r/R_0)$ vs $\log_{10}(v/c)$ for an open universe around $z \approx 1$, see Eqs. (18.11) and (18.12).

---

[b]In a large size "black hole", exploding or not, there is a singularity at the center of mass ($R < R_{\text{HL}} \approx ct_{\text{HL}} \approx \hbar/M_u c$) which is of the order of 10^{-98} cm for an open universe and ten times less for a flat universe. Only after $R > R_{\text{Sch}} = 2GM_u/c^2$ local gravitational attraction can be expected to give rise to stars and galaxies around local mass fluctuations. Before that it may be expected that gravitational attraction is centered in the center of mass of the "black hole".

[c]This is taken from Fig. 17.2 in Joseph Silk's *The Big Bang* (Third Edition) p. 366, which plots "apparent magnitude" against redshift for distant galaxies at $z \approx 1$, currently interpreted as "cosmic acceleration", which is remarkably similar to what appears in Fig. 18.1, due strictly to relativistic effects specified by Eqs. (18.16) and (18.17) existing even in a universe with $\Lambda = 0$.

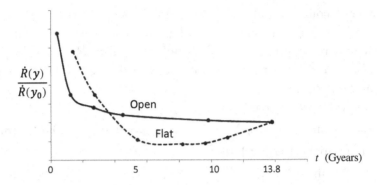

Fig. 18.2.　$\dot{R}(y)/\dot{R}(y_0)$ for an open (solid line) and a flat (dotted line) universe as a function of time.

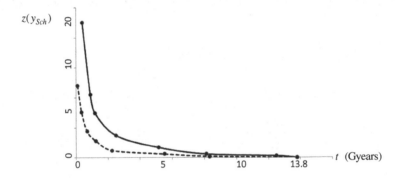

Fig. 18.3.　The redshift z for an open and a flat universe as a function of time showing that $(z_{obs})_{max}$ is compatible with the first and not with the second.

totally unrelated to any cosmic "acceleration" inexistent in the open universe with $\Lambda = 0$.

As shown in Ref. 5, taking into account the zero-point energy after decoupling atom formation ($T_{dec} \approx 2178$ K) in an open universe ($k < 0, \Lambda = 0$), energy conservation is justified because

$$\frac{\rho_r}{\rho_m} + \frac{\rho_{zp}}{\rho_m} = \frac{T}{T_{dec}} + \left(1 - \frac{T}{T_{dec}}\right) = 1, \qquad (18.20)$$

where ρ_r is the radiation density, ρ_{zp} the zero-point energy density and $T_{dec} = 2178$ K is the decoupling temperature. That is not the case in a flat ($k = 0, \Lambda > 0$) universe.

Table 18.2 gives quantitative comparisons of results obtained in an open and a flat universe, including those for the "last scattering" CMB spectrum, which are almost indistinguishable for $\Omega_m(\text{LS})$ and $H(\text{LS})t(\text{LS})$ and seem to be better for the flat model after laborious

Table 18.2. Quantitative comparisons of results obtained in an open and flat universe.

Open Universe (Energy Conservation)	Flat Universe (Energy Non-Conservation)
$t(y) = [R_+/c|k|^{1/2}][\sinh y \cosh y - y]$	$t(y) = \dfrac{2}{3}\left(\dfrac{\Lambda}{3}c^2\right)^{-1/2} y$
$R(y) = R_+ \sinh^2 y$	$R(y) = R(\Lambda)\sinh^{2/3} y$

$$t_0 = 13.8 \times 10^9 \text{ yr}, \; H_0 = 67.1 \text{ km/s/Mpc}$$

$H(y_0)t(y_0) = 0.940 \to y_0 = 2.241$; $\quad y = y_{Sch} = 0.8813$	$H(y_0)t(y_0) = 0.940 \to y_0 = 1.160$; $\quad y_{Sch} = 0.2442$
$\Omega_{m0} = 0.044$ (dark matter $= 0$)	$\Omega_{m0} = 0.322$ (dark matter $= 0.278$)
$[\langle\Omega_m\rangle]^{y_0}_{y_{Sch}} = 0.256$	$[\langle\Omega_m\rangle]^{y_0}_{y_{Sch}} = 0.616$
$\Omega_{k0} = 0.956$ ("dark energy")	$\Omega_{\Lambda0} = 0.678$ ("dark energy")
$M_u = \dfrac{c^2 R_{Sch}}{2G} = 4.105 \times 10^{54}$ g	$M_u = \dfrac{c^2 R_{Sch}}{2G} = 2.869 \times 10^{55}$ g
$T_{eq} = \dfrac{M_u/(T_0 R_0)^3}{\frac{4\pi}{3}\sigma/c^2} = 2178\,\text{K} \approx T_{af}$	$T_{eq} = \dfrac{M_u/(T_0 R_0)^3}{\frac{4\pi}{3}\sigma/c^2} = 15243\,\text{K} \gg T_{af}$

Cosmic acceleration at $z \approx 1$?

No	Yes
Using $v = c[(1+z)^2 - 1]/[(1+z)^2 + 1]$	Using $v = cz$

Very early galaxy formation at $t_{gf} \leq t_{Sch}$?

Yes	No
$t_0 - t_{Sch} \approx 13.6 \times 10^9$ yr	$t_0 - t_{Sch} \approx 10.9 \times 10^9$ yr

(*Continued*)

Table 18.2.   (*Continued*)

Open Universe (Energy Conservation)	Flat Universe (Energy Non-Conservation)
Inflation at $t_{\text{infl}} \approx 10^{-35}$ s?	
No $\rho(t_{\text{infl}} + \delta t) \approx \rho(t_{\text{infl}} - \delta t)$	Yes $\rho(t_{\text{infl}} + \delta t) \approx 10^{40}\rho(t_{\text{infl}} - \delta t)$
"Last scattering" CMB spectrum fit?	
Moderate $\Omega_m(\text{LS}) = 0.983$ $H(\text{LS})t(\text{LS}) = 0.668$ $l_{\max} \approx 190$	Good $\Omega_m(\text{LS}) = 0.999$ $H(\text{LS})t(\text{LS}) = 0.666$ $l_{\max} \approx 670$
Zero-point energy change at $T < T_{dec}$?	
Yes $\dfrac{\rho_r + \rho_{zp}}{\rho_m} = \dfrac{T}{T_{dec}} + \left(1 - \dfrac{T}{T_{dec}}\right)$	No $\dfrac{\rho_r + (0)}{\rho_m} = \dfrac{T}{T_{dec}}$

numerical optimization.[2] I would like to thank Manuel Alfonseca for helpful suggestions and competent criticism.

References

1. J. A. Gonzalo and M. Alfonseca, Constraints on the general solutions of Einstein's cosmological equations by Hubble parameter times cosmic age: A historical perspective, arXiv: 1306.0238; J. A. Gonzalo and M. Alfonseca, Is a single, finite, open Universe ruled out by Planck's satellite data? *The Bi-Monthly Journal of the BWW Society* **1** (2014) 13.
2. J. A. Gonzalo, *Cosmological Implications of Heisenberg's Principle* (World Scientific: Singapore, 2015), pp. 106–110.

Chapter 19

Singular Moments in Cosmic History

The Big Bang

A few microseconds after the beginning of space-time, about 13.7 billion years ago, baryons (neutrons) and high energy photons, at a temperature of $T \sim 4 \times 10^{12}$ K, were already expanding rapidly. A few minutes after, electrons were formed, and, in about 30 minutes, nucleosynthesis (mainly of ^{4}He nuclei) took place. The primordial ratio of neutrons-to-protons in the universe was fixed.

A few million years after, the expanding universe had cooled enough, to $T \sim 3 \times 10^3$ K, and atoms (mostly hydrogen but also substantial amounts ^{4}He) were formed. Matter and radiation decoupled and the universe became transparent.

After about 360 million years, stars and galaxies began to form.

The Sun and the Solar Planetary System

About five billion years ago, i.e., some nine billions years after the Big Bang, the Sun, a second or third generation star in the Milky Way, was formed out of cosmic dust from previous stars, which contained already substantial amounts of heavy elements from previous supernovae explosions. The solar system, made up by Mercury,

Venus, the Earth, Mars, Jupiter, Saturn, Uranus and Neptune, plus the asteroids, plus the comets and meteorites, was formed.

The Earth, located between Venus and Mars, occupied a privileged position, with a mass, a specific chemical composition and a magnetic field, which, in time, would allow our planet to develop a biosphere with plenty of water, and a surrounding gaseous atmosphere made up of oxygen and nitrogen.

The Earth–Moon System

The formation of planet Earth is dated to about 4 600 million years ago, and the formation of the Earth–Moon system, a lucky coincidence in all respects, is assumed to be the result of a cosmic clash between a very large foreign body and the primitive Earth. The Earth–Moon system is absolutely unique in our solar system. No other satellite has a mass comparable to that of the corresponding planet. This has important consequences: it stabilizes the rotation axis of the Earth and makes possible the four seasons during the year. It makes possible in our planet a night luminary for the most of the months. And, in due time, when the oceans and the Earth's atmosphere are formed, it makes the Moon the main agent of the tides, with important consequences.

The unique "habitat" made up by our *Earth–Moon system* depends on a number of physical *preconditions*.[1,2] In particular:

- The actual combination of specific universal *physical constants* which determines the scales of mass, size and time in the universe.
- The possibility of existence of stable and radioactive *nuclei* making up the Periodic Table of the elements.
- The possibility of existence of stable *atoms* and molecules at the temperatures on the Earth.
- The abundant availability of *light elements* (C, O, N, S) on the Earth's crust; heavy *non-metallic* elements (Mg, Ca, Str, Ba) in the mantle; and heavy *metallic* elements (Fe, Ni, Co) in the magnetic core, responsible among others things for the Earth's magnetic field.

- That the Earth is at the *right distance* from the Sun, such that its mean temperature is about 300 K, intermediate between the melting point of ice and the boiling point of water.
- That it has a *mass* and a *size* adequate to retain an atmosphere rich in O_2 and N_2 in the right proportion.

The list of physical and chemical preconditions could go on further. But this sample is enough to illustrate the differences with our two neighboring planets in the solar system: Venus and Mars. Kant and other eighteenth century enlightened philosophers expected intelligent beings in both. Today we know, through NASA's research satellites, that both are quite inhospitable to life.

Life

About 3 600 million years ago, the first fossil traces of life were left out in old carbonaceous rocks. The most elementary unicellular organism, a bacteria, has already a tremendous complexity. All living beings, from the bacteria to the most developed mammals, have a genetic code which, through its DNA, preordains its development, and through its RNA and its proteins, controls its metabolism, its immunologic defences and its reproductive system. The basic components of the cells constituting unicellular organisms, plants, and animals, are the same. And their genome is made up of long sequences of the same nucleotides: adenine, guanine, cytosine and thymine, linked by sugar and phosphate groups.

Francis Crick,[3] an agnostic, co-discoverer of the structure of DNA, and winner of the Medicine and Physiology Nobel Prize in 1962, says:

> *An honest man, armed with all the knowledge available to us now, could only state that in some sense, the origin of life appears at the moment to be almost a miracle, so many are the conditions which would have had to have been satisfied to get it going.*

A representative sample of these conditions[4] is the following:

- *Plate tectonics* and volcanic activity contributing to increase the H_2O content in the biosphere.

- An early, prolonged, *rain of meteorites* on the Earth's surface, contributing to the same end.
- The subsequent accumulation of *water*, with its exceptional physical and chemical properties, in the world's oceans.
- The extraordinary *chemical properties of C* which result in the possibility of organic chemistry.
- The exceptional combination of biochemical properties in the *amino acids*, the *proteins* and the *DNA macromolecules.*
- The marvel of *photosynthesis*, through which green plants take advantage of the solar flux during the day to synthesize organic matter, using CO_2 and liberating O_2.
- The subtle *biochemistry* that controls the metabolisms, the immunological system the reproductive mechanisms of the vertebrates.
- The fact that the Earth has a *large planetary companion* such as Jupiter, orbiting the Sun in the same plane, and protecting our planet from most asteroids and meteorites.
- The existence of a protective belt, the *Van Allen belt*, held in place by the Earth's magnetic field in the upper ionosphere, which protects our planet from too much cosmic radiation.
- The fortunate *location* of the Sun in an outer region of the *Via Lactea* in which the background galactic radiation is relatively low. As pointed out in *The Privileged Planet*,[4] the Sun's location and the Earth's location within the galaxy are ideal for a cosmic observatory, resulting in the possibility of observation of solar eclipses, and among other things, the testing of General Relativity.

Man

About 60 thousand years ago the last great glaciations took place. Some 15 thousand years ago, *primitive men*, with creative artistic skills comparable to those of Dalí or Picasso, painted bisons, horses and hunters in the walls of Altamira and Lascaux. Only 10 thousand years ago the inhabitants of primitive cities developed agriculture and domestication of animals (cattle, horses, camels, sheep).

There took place, as well, the first fabrication of ceramics and clay vessels.

It is *undeniable* that men share with the higher mammals, specially the apes, a great many physical characteristics. It is well known that the genome of the *chimpanzee* (living on Earth for more than five million years according to our contemporary anthropologists), differs from that of *men* only in *one percent*.

It is even more *undeniable* that men are quite different from ordinary apes:

- Men are the only apes that *speak*, in many languages (Spanish, English, Russian, Chinese...); the only ones that *write*, in many alphabets (Cuneiform, Roman, Cyrillic, Arabic...); the only ones that make *symbols*, like flags, logos, national anthems...
- Men are the only apes that truly appreciate the beauty of *sunset* or the splendor of a *starry night*; the only ones that have invented folk songs; and games, from football to chess; the only apes who have held Olympics Games, in ancient Greece and in our own twentieth century.
- Men have been the only apes able to built such large constructions as the *Great Chinese Wall* (it can be viewed by the astronauts from outer space), the *Egyptians Pyramids*, or the great medieval *European Cathedrals*.
- The only apes which have written *symphonies* and painted *masterpieces* such as the Sistine Chapel.
- The only ones which have written *epic poems* such as, the *Iliad*, or *novels* such as *Don Quixote*, or *dramas* such as *Hamlet*.
- The only apes which have *historical records*, or *calendars*, or *ships* to explore the five continents.
- Men are the only apes that have developed *scientific learning*: not only *grammar*, *logic* and *rhetoric*; but also *mathematics*: geometry and algebra; the only apes which, within the last three centuries, have developed *physics*: mechanics, electromagnetism, thermodynamics, and optics.
- Men are the only apes that have developed *advanced technologies*; can communicate at the speed of light; can travel in jets at

supersonic speeds; or have invented powerful computers, lasers and GPS.

- Men are the only apes — for the time being — which have been able to *fly to the Moon.*
- The only apes which through *special satellites* (Hubble, COBE, WMAP...) have been able to explore the last confines of the observable universe.
- It is true, however, that man is the only ape that has often forgotten that it is not exactly the same thing to penetrate the secrets of nature and to consider that he is the owner and lord of nature.

After putting together this rough sequence of singular events in cosmic history some may still insist that the sequence is a succession of *chance* occurrences.

Others, more realistic, would recognize something else: *order* and *purpose.*

References

1. M. Denton, *Evolution: A Theory in Crisis* (Adler and Adler, Bethesda, Maryland, 1986).
2. M. Denton, *Nature's Destiny* (The Free Press, New York, 1998).
3. F. Crick, *Life Itself* (Simon & Schuster, New York, 1981).
4. G. Gonzalez and J. W. Richards, *The Privileged Planet* (Regnery, Washington, 2004), Chapter 10.

Chapter 20

A Brief Outline: World Events and Cosmological Discoveries from −4500 to 2010

c.−4500:	Egyptian calendar.
c.−3000:	Astronomical observations in Egypt, Babylonia, India and China.
c.−2500:	Cheops pyramid constructed.
c.−2000:	Equinoxes and solstices determined in China.
c.−2000:	Abraham, Isaac, Jacob: Promise of the Messiah.
c.−1500:	Moses: The Ten Commandments.
c.−1500:	Advanced shipbuilding in Mediterranean and North Sea.
c.−800:	Babylonian and Chinese planetary tables.
c.−550:	Thales of Miletus predicts solar eclipse.
−480:	Buddha dies: His teaching spreads through India and all Asia.
−479:	Confucius dies: His teaching spreads through China and Korea.
−427/−323:	Plato (−427, −347), Aristotle (−384, −322), Alexander the Great (−356, −323)
−323:	Euclid's *Elements*

−250:   Aristarchus of Samos: First heliocentric proposal.
−240:   Erathostenes: Accurate estimate of radius of
      Earth.
−140:   Hipparchus: Improved astronomical observations.
      Discovery of the star "Novae". Trigonometry.
−46:   Adoption of Julian calendar in Roman Empire.
−7:   Probable date of Jesus Christ's birth in
      Bethlehem.

33:   Probable date of Jesus Christ's crucifixion in
      Jerusalem.
64:   Peter and Paul executed in Rome: Church
      spreads through Roman Empire.
94:   Trajan: Roman Empire reaches greatest
      geographic extension.
170:   Ptolemy: Geocentric planetary system (deferents
      and epicycles).
600:   Book printing in China.
622:   The Hijrah: Muhammad's flight from Mecca to
      Medina.
      Origin of Muslim calendar. Islam spreads through
      East and West.
828:   *Almagest*: Translation of Ptolemy's planetary
      tables to Arabic.
975:   Arabic arithmetic notation introduced in
      Medieval Europe.
1099:   First Crusade: Christians conquer Jerusalem.
1100–1300:   Medieval European Universities: Bologna (1119),
      Paris (1150), Oxford (1167), Salamanca
      (1217), Cambridge (1200) *etc.*
1200:   Saladin's army recovers Jerusalem.
1211:   Genghis Khan invades China.
**1330:   Jean Buridan (1295–1358) teaches
      "impetus" theory in Paris.
      Inertial motion.**
1450:   Fall of Constantinople to the Turks.

1492: Columbus's first voyage to New World.

1498: Vasco da Gama: Sea route to India surrounding Africa.

1500: Juan de la Cosa: First World Map.

1522: Magellan and Elcano: First circumnavigation of the world.

1529–1533: Division of Christendom: Luther (1483–1546), Zwingli (1484–1531), Calvin (1509–1564) break with Church.

1543: Copernicus: "De revolutionibus orbium celestium". Heliocentric proposal for planetary system.

1570: Battle of Lepanto.

1583: Adoption of Gregorian Calendar under Pope Gregory XIII.

1598: Tycho Brahe: Accurate measurements of planetary motions.

1610: Galileo Telescope. "Nuntius Sidereus". Copernican planetary system defended.

1618: Kepler: "Harmonices Mundi". Laws of planetary motion.

1675: Römer: Measurement of finite speed of light.

1687: Newton "Principia". Laws of inertial motion. Gravitation.

1764: James Watt: Invention of heat engine.

1780: American revolution.

1789: French revolution.

1820: W. Herschel: Catalog of 5000 nebulae in the northern hemisphere.

1829: Non-Euclidean geometries: Gauss (1777–1855), Lobachevski (1793–1856), Bolyai (1802–1860).

1836: F. W. Bessel: Parallax of 61 Cygni measured.

1842: Mayer: Principle of conservation of energy.

1873: Maxwell: Maxwell's equations.

1900: Planck: Black body radiation law.

1914–1918: World War One.

1917:	Russian revolution.
1916:	**Einstein's General Relativity. Cosmological equations.**
1922:	**Friedmann's solutions to Einstein's equations.**
1927:	**Lemaitre's solutions to Einstein's equations (independently) Big Bang model for a expanding universe.**
1939–1945:	**World War Two.**
1948:	**Alpher and Herman predict cosmic background radiation (CBR).**
1965:	**Penzias and Wilson detect CBR.**
1989:	**COBE: Planck's spectral distribution of CBR measured (Mather). Minute anisotropies in CBR measured (Smoot).**
2003:	**WMAP: Improved measurements. "Age" of universe precisely estimated: 13.7 $\times 10^9$ yrs. Big Bang basically confirmed.**

We may conclude this section and this book with the following eloquent quotation:

The system to which we have likened the universe is an intelligible system. It contains no irrationalities save the one supreme irrationality of creation — an irrationality indeed to physics, but not necessarily to metaphysics... Those who feel the question ["Why the universe?"] to be a permissible one, can legitimately answer the question "why?" by positing God.

The physicist and cosmologist then need God only once, to ensure creation. For the biologist the world provides further opportunity for divine planning, if we admit the possibility of not entirely coincident evolutionary trends in similar circumstances. For man as more than cosmologist, as more than biologist, as possessing mind, possibly endowed with an immortal soul, God is perhaps needed always. Theoretical cosmology is but the starting-point for deeper philosophical inquiries".

E. A. Milne, in *Relativity, Gravitation and World Structure* (Clarendon Press, Oxford, 1935), p. 140.

Notes Added in Production

The latest data from NASA's "James Webb Space Telescope" (JWST) given in July 12, 2022, reveal a vastly improved look at the deep field image seen by the "Hubble Telescope" long time ago. The JWST is a 10 billion telescope looking at the sky in the infrared, cooled at $-233°C$ to improve the accuracy, put in the second Lagrange point at 1.5×10^9 kms from Earth. The mirror is veneered in gold is 6.5 m in diameter, the largest sent to space. It is designed to be functional for twenty years. The image announced by NASA is the most precise further way cosmic image located at 4.6×10^9 light years from Earth. The JWST data show images of giant planets, dying stars and galaxies striking each other at very high speeds. It is too early to imagine the extremely valuable cosmic data which the JWST will provide in the near future.

Appendix A

Constraints on the General Solutions of Einstein's Cosmological Equations by Hubble Parameter Times Cosmic Age: A Historical Perspective

Julio A. Gonzalo* and Manuel Alfonseca†

*Escuela Politécnica Superior, Universidad San Pablo CEU,
Montepríncipe, Bohadilla del Monte, 28668 Madrid, Spain
Departamento de Física de Materiales, Universidad
Autónoma de Madrid, 28049, Madrid, Spain
†Escuela Politécnica Superior, Universidad San Pablo CEU,
Universidad Autónoma de Madrid, 28049, Madrid, Spain

In a historical perspective, compact solutions of Einstein's equations, including the cosmological constant and the curvature terms, are obtained, starting from two recent observational estimates of the Hubble's parameter (H_0) and the "age" of the universe (t_0). Cosmological implications for ΛCDM (Λ Cold Dark Matter), KOFL (k open Friedmann–Lemaitre), plus two mixed solutions are investigated, under the constraints imposed by the relatively narrow current uncertainties. Quantitative results obtained for the KOFL case seem to be compatible with matter density and the highest observed redshifts from distant galaxies, while those obtained for the ΛCDM may be more difficult to reconcile.

Keywords: Cosmic dynamical equations; flat cosmic solutions; open cosmic solutions; intermediate cosmic solutions.

---

†Corresponding author.

1. Introduction: Relevance of the Dimensionless Product $H_0 t_0$

In the late 70's, large uncertainties surrounded the numerical estimates of Hubble's ratio ($H_0 = \dot{R}_0/R_0$) and the "age" of the universe. The uncertainty in the density parameter ($\Omega = \rho_0/\rho_{co}$) giving the ratio between the current density and the critical density, $\rho_{co} = 3H_0^2/8\pi G$, was also very large.

An attempt was made by Beatriz Tinsley[1] to explore what type of universe we live in, with the scarce information then available. She pointed out then that the universe was most likely open, assuming a zero cosmological constant ($\Lambda = 0$) in Einstein's cosmological equations. Furthermore, she noted that the dimensionless product $H_0 t_0$, which requires only local data, can be especially advantageous to characterize the cosmic equation.

At that time, many cosmologists stated their preference for a closed universe. They hoped that the amount of ordinary matter in the universe would be proven large enough to close the universe, but not by much, since otherwise the expansion would have been reversed by this time, something that obviously has not happened.

In the nineties, at a Summer Course on Astrophysical Cosmology[2] which took place in El Escorial, Spain, a number of distinguished experts were present, including Ralph Alpher, John C. Mather, George F. Smoot, Hans Elsässer and Stanley L. Jaki. The general consensus in the years before that time was that H_0 should be somewhere between 50 and 100 km/s/Mpc, and t_0 somewhere between 10 and 20 Gyrs.

The same year, the Harvard's group[3] offered a new, far better estimation, obtained by analyzing type Ia supernovae in far galaxies. H_0 was estimated at 67 km/s/Mpc and t_0 at 13.7 Gyrs.

Today the old uncertainties have come down sharply. New accurate estimates have been given for H_0 and t_0. In particular, those obtained from the Wilkinson Microwave Anisotropy Prove (WMAP) and published in December 2012 give[4,5]:

$$H_0 = 69.3 \pm 1.8 \text{ km s}^{-1}\text{Mpc}^{-1} = 2.246 \times 10^{-18}\text{s}^{-1} \quad (A.1)$$

$$t_0 = 13.77 \pm 0.11 \text{ Gyr} = 4.345 \times 10^{17}\text{s}. \quad (A.2)$$

while those published in March 2013 as a result of the analysis of the Planck telescope data give[6,7]:

$$H_0 = 67.15 \pm 1.2 \text{ km s}^{-1}\text{Mpc}^{-1} = 2.176 \times 10^{-18}\text{s}^{-1} \quad (A.3)$$

$$t_0 = 13.798 \pm 0.037 \text{ Gyr} = 4.354 \times 10^{17}\text{s}. \quad (A.4)$$

Curiously enough, the latest estimation for H_0 goes back to the value given in 1995 by Harvard's group. We have decided to consider both alternatives.

These data result in the dimensionless products $H_0 t_0 = 0.9759$ and $H_0 t_0 = 0.9476$ respectively, both smaller than, but relatively close to unity.

During the eighties, better and better measurements made clear that the amount of ordinary matter is much smaller (about 5%) than would be needed for a flat or a marginally closed universe. This, and some irregularities in the rotation of galaxies, brought to the conclusion that there must exist another kind of matter (dark matter) that would close the universe or at least make it flat. All the searches during the next thirty years failed to underpin dark matter, but the current estimations are quite small (about 27% of the critical value).

In 1998, a further analysis of type Ia supernovae by the Harvard group gave the (then) unexpected result that the universe is accelerating.[8] This gave rise to the following consequences:

- The cosmological constant in Einstein's cosmological equations (see below) was resurrected. For several decades, its value had been assumed to be zero.
- A new mysterious entity (dark energy) was introduced to represent the effect of the cosmological constant. Its effect (currently unexplained) would be the same as a negative gravity, giving rise to the currently accelerated expansion of the universe.

- The flat model of the universe became the standard cosmological assumption. The amount of dark energy was estimated as precisely what was needed to make that model possible (about 68% of the total mass of the universe). The closed model has been abandoned, as it is incompatible with the acceleration. Research on the possibility of an open model with a non-zero curvature is now also neglected.

However, even at the time when the acceleration effect was discovered, other reasons apart from dark energy were suggested, which could explain at least a part of the effect. Brightness attenuation due to early substantially denser cosmic dust[9] was offered as a possible explanation. Another one is proposed here.

Galaxies recede from the cosmic center of mass (assumed to coincide with the center of the expanding Cosmic Microwave Background Radiation sphere). A galaxy at a distance $r = R_0 - R$ from the Milky Way galaxy (which moves at $v_0 \approx 2 \times 10^{-3}c$ from the center of the CMBR sphere), recedes from us with a velocity v. These two variables, r and v, are characterized by the redshift z of that galaxy in this way:

$$v = c\frac{(1+z)^2 - 1}{(1+z)^2 + 1}, \tag{A.5}$$

$$r = R_0 \frac{z}{1+z}. \tag{A.6}$$

This means that, for $z \ll 1$, both v/c and r/R_0 approach z. Here R_0 is the radius of the expanding observable universe, which is greater than the radius of the CMBR sphere, but close to it at present, since the CMBR is receding from the Milky Way at $z \approx 1089$, which corresponds[9] to a velocity very close to c.

Figure A.1 depicts $\log_{10}(r/R_0)$ versus $\log_{10}(v/c)$ as given respectively by Eqs. (A.5) and (A.6), for z varying from 0.1 to 100. It shows that for $z > 1$ (see marked points for $z = 0.5$ to 10) the assumed proportionality between distance (r) and recession velocity (v) is no longer valid. Of course, for small values of z the proportionality is almost perfect. It should be noticed that observed values as large as $z = 10$ have been reported for distant galaxies. The upward trend of distance (or equivalently magnitude) vs. recession speed (velocity)

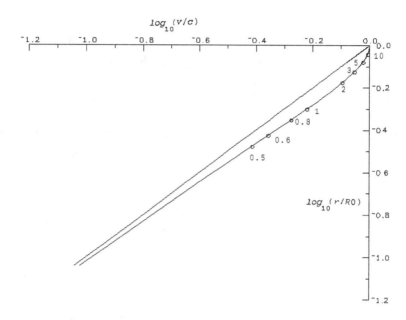

Fig. A.1. Distance/velocity curve parametrized by redshift z.

in Fig. A.1 mimics what is to be expected in an accelerated expansion, but is not related to a non-zero cosmological constant. So, it may explain, at least in part, the upward trend in magnitude vs. redshift observed in the case of type Ia supernovae, for $z > 1$ (the effect discovered in 1998).

2. Parametric Solutions for Einstein's Universe

Our starting point is the finite universe described by Einstein's cosmological equation[10] which can be written as

$$\dot{R}^2 = \frac{2GM}{R} - kc^2 + \frac{\Lambda}{3} c^2 R^2, \qquad (A.7)$$

where $\dot{R}$ is the time derivative of the radius of the observable universe R, $G = 6.67 \times 10^{-11}$ IS units, M the finite mass of the observable universe, k the space-time curvature ($k < 0$ for an open universe), $c = 3 \times 10^8$ m/s the speed of light, and Λ (cm^{-2}) is Einstein's cosmological constant, originally introduced by Einstein to counter gravitation.

Compact parametric solutions of Eq. (A.7) can be obtained easily in terms of $t(y)$ and $R(y)$, where y is a cosmic parameter going from $y \ll 1$ just after the singularity ($t = 0$, $R = 0$), to $y \gg 1$ well away from it. Explicit analytical solutions can be obtained for (i) a **flat universe** ($k = 0$, $\Lambda > 0$) (not previously reported in compact parametric form, as far as we know); (ii) an **open universe** ($k < 0$, $\Lambda = 0$), equivalent to the well-known Friedmann–Lemaître solution. There is also a third case, (iii) a **mixed universe** ($k < 0$, $\Lambda > 0$), where analytical solutions are not available, but the equations can be solved numerically.

In each case, from $t(y)$ and $R(y)$, $\dot{R}(y)$, etc., can be derived a simple dimensionless expressions for $H_0(y)t_0(y)$ and $\Omega_0(y)$, which will be shown useful to make direct quantitative comparisons with observational data, putting stringent constraints on the validity of the respective solutions.

2.1. *Parametric solutions for a flat ΛCDM universe*

With $k = 0$, $\Lambda > 0$, Eq. (A.7) results in:

$$\int_0^t dt = \int_0^R \frac{R^{1/2}}{\left(2GM + \left(\frac{\Lambda}{3}c^2\right)R^3\right)^{1/2}} dR$$

$$= \frac{2/3}{\left(\frac{\Lambda}{3}c^2\right)^{1/2}} \int_0^x \frac{1}{(a^2 + x^2)^{1/2}} dx \qquad (A.8)$$

where $x^2 \equiv \left(\frac{\Lambda}{3}c^2\right)R^3$, $a^2 \equiv \left(\frac{\Lambda}{3}c^2\right)R^3(\Lambda)$, which implies $R(\Lambda) = (2GM/\frac{\Lambda}{3}c^2)^{1/3}$.

This integral can be solved analytically using the change of variable $\frac{x}{a} = \sinh y$ and gives

$$t = \frac{2}{3}\left(\frac{\Lambda}{3}c^2\right)^{-1/2} y, \quad R = R(\Lambda)\sinh^{2/3} y, \quad \Lambda = \frac{4y^2}{3t^2c^2}. \qquad (A.9)$$

It can be seen that Λ, as a function of time, is completely determined by Einstein's equation in this model. Therefore, its current value (Λ_0) depends exclusively on y_0 and t_0, or, in other words, on the current estimations of H_0 and t_0 (see Table A.1 on page 137).

Table A.1. Cosmic parameters for a flat (ΛCDM) universe. Left, $H_0 = 69.3$ km/s/Mpc, $t_0 = 13.77$ Gyr, $\Lambda_0 = 1.2001e\text{-}52$ m^{-2}, $M_0 = 2.5811e52$ kg. Right, $H_0 = 67.15$ km/s/Mpc, $t_0 = 13.798$ Gyr, $\Lambda_0 = 1.0764e\text{-}52$ m^{-2}, $M_0 = 2.9595e52$ kg. In both cases, $T_0 = 2.72548$ K.

| | WMAP | | | | | Planck | | | | |
Radius	R (Mly)	y	t(My)	z	Ω_m	R (Mly)	y	t(My)	z	Ω_m
R_0	14110	1.2359	13770	0	0.287	14562	1.1729	13798	0	0.319
R_Λ	10421	0.8814	9820	0.354	0.5	11310	0.8814	10369	0.2875	0.5
R_{Sch}	4052	0.2401	2676	2.482	0.944	4646	0.2603	3063	2.134	0.935
R_{CMBR}	12.82	4.3e-5	0.48	1099.7	1	13.23	4e-5	0.47	1099.7	1

We can proceed now to get the relevant cosmic parameters. For the speed at which the cosmic radius is growing we get

$$\dot{R}(y) = \frac{dR/dy}{dt/dy} = R(\Lambda)\left(\frac{\Lambda}{3}c^2\right)^{1/2}\sinh^{-1/3} y \cosh y. \qquad (A.10)$$

Hubble's parameter then becomes

$$H(y) = \frac{\dot{R}(y)}{R(y)} = \left(\frac{\Lambda}{3}c^2\right)^{1/2}\frac{\cosh y}{\sinh y} \qquad (A.11)$$

and $H(y)t(y)$, using Eqs. (A.9) and (A.11), results in

$$H(y)t(y) = \frac{2}{3}\frac{y}{\tanh y}, \qquad (A.12)$$

which goes from $H(0)t(0) = 2/3$ to $H(y)t(y)$ growing indefinitely for $y \gg 1$.

The dimensionless density parameter $\Omega(y)$ becomes

$$\Omega(y) = \frac{\rho(y)}{\rho_c(y)} = \frac{M/\frac{4\pi}{3}R^3(y)}{3H^2(y)/8\pi G} = 1 - \tanh^2 y. \qquad (A.13)$$

Table A.1 shows some of the results obtained for the two boundary conditions we have used: Eqs. (A.1), (A.2) on the one hand, and Eqs. (A.3), (A.4) on the other. In this case, the differences between the two sets of cosmic parameters are not too high. The time of the CMB radiation is nearest to the number usually given (370,000 years after the Big Bang): for both scenarios, the times computed are 470,000 and 480,000 years, respectively. But the value of z when the radius of the observable universe was equal to the Schwarzschild radius for the computed mass of the observable universe seems too low, as will be explained later. Such values would have the consequence that most of the far, currently visible galaxies would have started to form when the whole universe was still an exploding black hole.

2.2. *Parametric solutions for an open (KOFL) universe*

Equation (A.7), with $k < 0$, $\Lambda = 0$, results in:

$$\int_0^t dt = \int_0^R \frac{R^{1/2}}{(2GM + c^2|k|R)^{1/2}} dR$$

$$= \frac{2}{c^3|k|^{3/2}} \int_0^x \frac{x^2}{(a^2 + x^2)^{1/2}} dx, \qquad (A.14)$$

where $x^2 \equiv c^2|k|R$, $a^2 \equiv c^2|k|R_+$, which implies $R_+ = 2GM/|k|c^2$.

The integral can be solved analytically using the change of variable $\frac{x}{a} = \sinh y$ and gives

$$t = \frac{R_+}{c|k|^{1/2}}[\sinh y \cosh y - y], \quad R = R_+ \sinh^2 y. \qquad (A.15)$$

We can proceed now to get the relevant cosmic parameters.

For the speed at which the cosmic radius is growing we get

$$\dot{R}(y) = \frac{dR(y)/dy}{dt(y)/dy} = |k|^{1/2} c \frac{1}{\tanh y}. \qquad (A.16)$$

Hubble's parameter then becomes

$$H(y) = \frac{\dot{R}(y)}{R(y)} = \frac{|k|^{1/2} c}{R_+} \frac{\cosh y}{\sinh^3 y} \qquad (A.17)$$

and $H(y)t(y)$, using Eqs. (A.15) and (A.17), results in

$$H(y)t(y) = \frac{1}{\tanh^2 y} - \frac{y}{\tanh y \sinh^2 y}, \qquad (A.18)$$

which goes from $H(0)t(0) = 2/3$ to $H(y)t(y) = 1$ for $y \gg 1$.

Table A.2. Cosmic parameters for an open (KOFL) universe, $k = -1$. Left, $H_0 = 69.3$ km/s/Mpc, $t_0 = 13.77$ Gyr, $M_0 = 1.1479$e51 kg. Right, $H_0 = 67.15$ km/s/Mpc, $t_0 = 13.798$ Gyr, $M_0 = 3.585$e51 kg. In both cases, $T_0 = 2.72548$ K.

	WMAP					Planck				
Radius	R (Mly)	y	t(My)	z	Ω_m	R (Mly)	y	t(My)	z	Ω_m
R_0	14199	2.8797	13770	0	0.0125	14835	2.3384	13798	0	0.0365
$R_+ = R_{\mathrm{Sch}}$	180.20	0.8814	96.0	77.8	0.5	562.78	0.8814	299.9	25.36	0.5
R_{CMBR}	12.90	0.2645	2.253	1099.7	0.933	13.48	0.1541	1.381	1099.7	0.977

The dimensionless density parameter $\Omega(y) = \rho_m/\rho_{mc}$ (which in fact corresponds to Ω_m) becomes

$$\Omega(y) = \frac{\rho(y)}{\rho_c(y)} = 1 - \tanh^2 y. \tag{A.19}$$

This analytical expression is identical to the one derived for the open universe, although one must remember that the meaning of parameter y is different in the two situations.

Table A.2 shows some of the results obtained for the two boundary conditions we have used: Eqs. (A.1), (A.2) on the one hand, and Eqs. (A.3), (A.4) on the other. In this case, the differences between the two sets of cosmic parameters are higher than in the flat universe, especially regarding the Schwarzschild radius and time, which are over three times larger for a slightly smaller value of H_0. This means that this model is very sensitive to the actual value of H_0. In both cases, however, the corresponding value of z is higher than the currently observed farthest galaxy, which means that all of them were formed when the observable universe was not an exploding black hole. The time of the formation of the CMB radiation is significantly higher than in the flat case: 1,381,000 and 2,253,000 years, respectively.

Table A.3 compares some of the results obtained for the open (KOFL universe) with several values of the curvature k. It can be seen that the time computed for the formation of the CMB radiation does not depend on the value of k. The radius of the observable universe and the Schwarzschild radius, however, get quickly smaller when k goes to zero. In fact, the first radius is smaller than the distance traveled by light since the CMB radiation formed, for all

Table A.3.   Results comparison for an open (KOFL) universe with different values of k. Left, $H_0 = 69.3$ km/s/Mpc, $t_0 = 13.77$ Gyr, $M_0 = 1.1479e51$ kg. Right, $H_0 = 67.15$ km/s/Mpc, $t_0 = 13.798$ Gyr, $M_0 = 3.585e51$ kg. In both cases, $T_0 = 2.72548$ K.

| | WMAP | | | | Planck | | | |
k	R (Mly)	R_{Sch} (Mly)	z_{Sch}	t_{CMBR}	R (Mly)	R_{Sch} (Mly)	z_{Sch}	t_{CMBR}
-1	14199	180.2	77.8	2.253	14835	562.8	25.36	1.381
-0.9	13470	153.9	86.6	2.253	14074	480.5	28.29	1.381
-0.75	12297	117.0	104.1	2.253	12848	365.5	34.15	1.381
-0.5	10040	63.7	156.6	2.253	10490	199.0	51.72	1.381
-0.1	4490	5.7	787.0	2.253	4691	17.8	262.6	1.381

$k \geq -0.9$ in the first scenario, and for all $k > -0.9$ in the second scenario.

2.3.   *Numerical solutions for a mixed universe*

The numerical solutions for a mixed universe ($k < 0$, $\Lambda > 0$) can be easily obtained by solving the Einstein's equation (A.7) for the appropriate value of the mass of the observable universe. As this value is unknown, this parameter must be adjusted by successive approximations. The Einstein's equation cannot be solved numerically starting at $t = 0$ (the Big Bang itself), because at that point there is a singularity, therefore we decided to start solving the equation at $t = t_{CMBR}$, which means that the initial condition (the time at which this phenomenon took place) must also be estimated. We did it by interpolating between the corresponding values for the flat universe solution and the open universe solution. Table A.4 shows the results for $k = -0.5$, $\Lambda = \Lambda_0/2$, $M_0 = 1.375 \cdot 10^{52} - 1.9114 \cdot 10^{52}$ kg and $k = -0.75$, $\Lambda = \Lambda_0/4$, $M_0 = 5.25 \cdot 10^{51} - 9.8 \cdot 10^{51}$ kg (using the value of Λ_0 for each scenario).

Looking at Table A.4, it can be seen that only the left lower part is compatible with the formation of all visible galaxies after the observable universe stopped being an exploding black hole. The origin

Table A.4. Cosmic parameters for a mixed universe. Left, $H_0 = 69.3$ km/s/Mpc, $t_0 = 13.77$ Gyr. Right, $H_0 = 67.15$ km/s/Mpc, $t_0 = 13.798$ Gyr. In both cases, $T_0 = 2.72548$ K.

Radius ($\Lambda = \Lambda_0/2$, $k = -0.5$)	WMAP				Planck			
	R (Mly)	t (Mly)	z	Ω_m	R (Mly)	t (Mly)	z	Ω_m
R_0	13833	13770	0	0.124	15409	13798	0	0.174
R_{Sch}	1648	967.3	7.4	0.665	3000	1758.5	4.14	0.660
R_{CMBR}	12.567	0.93	1099.7	0.996	14.0	0.92	1099.7	0.9977

Radius ($\Lambda = \Lambda_0/4$, $k = -0.75$)	WMAP				Planck			
	R (Mly)	t (Mly)	z	Ω_m	R (Mly)	t (Mly)	z	Ω_m
R_0	13989	13770	0	0.0599	14940	13798	0	0.0993
R_{Sch}	824	459.8	15.97	0.571	1538	858.2	8.71	0.571
R_{CMBR}	12.709	1.15	1099.7	0.9886	13.57	1.15	1099.7	0.9934

of the CMB radiation in these cases would have happened around one million years after the Big Bang.

3. Constraints on the Time Dependence of the Density Parameter Ω_m

Figure A.2 shows $\Omega(y)$ vs. $H(y)t(y)$ for a **flat** universe, Eq. (A.13), and an **open** universe, Eq. (A.19), using the Planck data. Note that the current situation ($H_0 t_0$) is represented in both models by the same abscissa (the dotted line), while the half-density parameter $\Omega(y) = 1/2$, signaled by separate arrows for both models, is substantially different (see Tables A.1 and A.2). $H(y)t(y) < 2/3$ corresponds to closed universes, currently discarded. For the open universe (the line below, as indicated in the figure), $2/3 < H(y)t(y) < 1$; for the flat universe (the line above), $2/3 < H(y)t(y) < \infty$. The shaded section would correspond to the whole range of different mixed cases.

Equation (A.7) can be rewritten[9] as:

$$1 = \Omega_m + \Omega_k + \Omega_\Lambda \qquad (A.20)$$

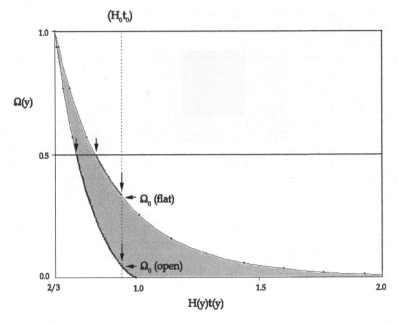

Fig. A.2.   Matter density parameter $\Omega(y)$ vs. dimensionless cosmic parameter $H(y)t(y)$ = Hubble's ratio × time for an open (OFLM) and a flat (ΛCDM) universe.

where Ω_m is the time-dependent mass density parameter (actually matter mass plus radiation mass, although at present radiation density is much less than matter density), Ω_k is the time-dependent space-time curvature energy density (which equals zero if $k = 0$), and Ω_Λ is the time-dependent energy density associated to Λ (which equals zero if $\Lambda = 0$). This relationship holds at present and at any time since the Big Bang.

Figure A.3(a), for the flat (ΛCDM) model, displays the evolution of $\Omega = \Omega_m + \Omega_r$ and Ω_Λ (due to the cosmological constant or "dark energy") as a function of z (obviously related to $T(K)$, t (gyrs), y (dimensionless), from the cosmic microwave background radiation to $z = 0$. The dotted lines correspond to t_{Sch} (3.063 Gyrs) and t_0 (13.798 Gyrs). Figure A.3(b) displays the evolution of $\Omega = \Omega_m + \Omega_r$ and Ω_k as a function of z, for the open (OFLM) model, where the dotted lines correspond to t_{Sch} (0.2999 Gyrs) and t_0 (13.798 Gyrs). In both cases, the Planck data are used.

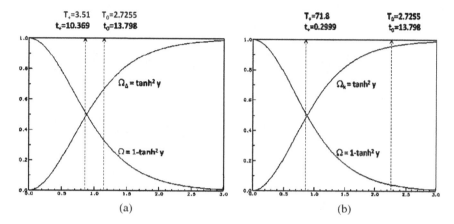

Fig. A.3. (a) Ω and Ω_Λ vs. z for a flat (ΛCDM) model. (b) Ω and Ω_k vs. z for an open (KOFL) model.

When the radius of the universe grew beyond the Schwarzschild radius $R_{\text{Sch}} = 2GM/c^2$ (this happened long after baryon formation, nucleosynthesis and atom formation, and the decoupling of the cosmic microwave background radiation) the universe was already transparent. We can assume that galaxies did not form before that time (when the behavior of the universe would have been similar to that of an exploding black hole), but started forming soon after. Sometime later, when galaxies and stars were fully formed, they began to emit redshifted light, which is now arriving to us. At present:

$$1 = \Omega_{m0} + \Omega_{k0} + \Omega_{\Lambda 0}. \tag{A.21}$$

And averaging from the time of galaxy formation to the present we have:

$$1 = \langle \Omega_m \rangle + \langle \Omega_k \rangle + \langle \Omega_\Lambda \rangle. \tag{A.22}$$

This is illustrated in Fig. A.4, which has been drawn using the data in Eqs. (A.3) and (A.4), for the flat ΛCDM and for the open KOFL universe. A quantitative discussion will take place in Sec. 4 for the cases corresponding to Eqs. (A.1), (A.2) and (A.3), (A.4). It must be noted that in Fig. A.4 the respective values of Ω_m and Ω_Λ or Ω_k are time-dependent from the Big Bang ($\Omega_m = 1$) to time growing indefinitely ($\Omega_m \to 0$) as required by Eq. (A.13).

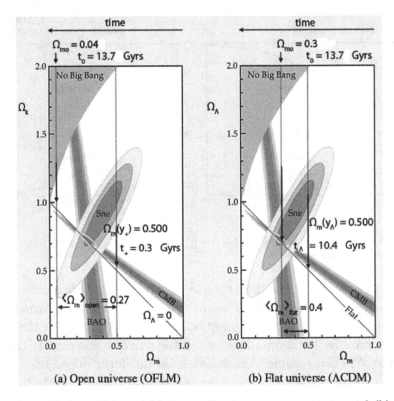

(a) Open universe (OFLM)   (b) Flat universe (ACDM)

Fig. A.4.   Time evolution of (a) Ω_k vs. Ω_m for an open universe, and (b) Ω_Λ vs. Ω_m for a flat universe. Confidence contours from the cosmic microwave background (CMB) and other constraints are shown. (See *Physics Today*, Dec 2011, pp. 14–17)

In summary, we have shown that, as anticipated by Beatriz Tinsley in the late 70s, a judicious use of the dimensionless product $H(y)t(y)$ with properly compact solutions of Einstein's cosmological equation, lead to a possible discrimination between the two main cosmological models discussed here.

4. Quantitative Comparisons between Flat ΛCDM, Open KOFL and Mixed Solutions

Tables A.5 and A.6 show the values predicted for several cosmic quantities for both the flat ΛCDM and the open KOFL solutions in two

Table A.5. Cosmic parameters for a flat (ΛCDM), mixed and open (KOFL) universe with WMAP results: $H_0 = 69.3$ km/s/Mpc, $t_0 = 13.77$ Gyr, $T_0 = 2.726$ K.

Model	y_0	M_u (10^{51} kg)	R_0 (Mly)	T_{Sch} (K)	z_{Sch}	Ω_{m0}	$\langle\Omega_m\rangle$	$\langle\Omega_x\rangle$
ΛCDM ($\Lambda_0\rangle$ 0)	1.2359	25.81	14110	9.49	2.482	0.287	0.616	0.384
KOFL ($k = -1$)	2.8797	1.148	14199	214.8	77.8	0.0125	0.256	0.744
Mixed ($\Lambda_0/2$, $k = -0.5$)		13.75	13833	22.87	7.39	0.124	0.394	0.606
Mixed ($\Lambda_0/4$, $k = -0.75$)		5.25	13989	46.26	15.97	0.0599	0.315	0.685

Table A.6. Cosmic parameters for a flat (ΛCDM), mixed and open (KOFL) universe with Planck results: $H_0 = 67.15$ km/s/Mpc, $t_0 = 13.798$ Gyr, $T_0 = 2.726$ K.

Model	y_0	M_u (10^{51} kg)	R_0 (Mly)	T_{Sch} (K)	z_{Sch}	Ω_{m0}	$\langle\Omega_m\rangle$	$\langle\Omega_x\rangle$
ΛCDM ($\Lambda\rangle$ 0)	1.1729	29.60	14562	8.54	2.134	0.319	0.627	0.373
KOFL ($k = -1$)	2.3384	3.585	14835	71.8	25.36	0.0365	0.268	0.732
Mixed ($\Lambda_0/2$, $k = -0.5$)		19.114	15409	14.0	4.14	0.174	0.417	0.583
Mixed ($\Lambda_0/4$, $k = -0.75$)		9.8	14940	26.47	8.71	0.0993	0.335	0.665

different cases:

- $H_0 = 69.3$ km/s/Mpc, $t_0 = 13.77$ Gyr and
- $H_0 = 67.15$ km/s/Mpc, $t_0 = 13.798$ Gyr.

In both cases, y_0 has been determined from the dimensionless value of $H_0 t_0$, and the remaining values have been obtained thus:

- $R_0 = c/H_0$
- $R_{Sch} = R[y_{Sch} = \sinh^{-1}(1)]$
- $T_{Sch} = R_0 T_0 / R_{Sch}$
- $z_{Sch} = T_{Sch}/T_0 - 1$
- $\Omega_{m0} = 1 - \tanh^2(y_0)$
- $\langle \Omega_m \rangle = (\Omega_{m0} + \Omega_{Sch})/2$
- $\langle \Omega_x \rangle = 1 - \langle \Omega_m \rangle$ with $x = \Lambda, k$.

Two additional mixed cases, computed numerically, have been added to the tables.

It can be seen that the values for the Schwarzschild redshift and the densities for the open (KOFL) universe are in better agreement with present observational expectations, especially in Table A.6. In both tables, the flat universe would be untenable, since the maximum observable redshift for galaxies (z_{Sch}) would appear to be significantly less than $z \approx 10$, currently observed.

5. Concluding Remarks

We have shown that, as pointed out long ago by Beatriz Tinsley, using the dimensionless product $H(y)t(y)$ in conjunction with the dimensionless density ratio $\Omega_m(y)$, imposes stringent constraints on the solutions of Einstein's cosmological equation, pointing to a better understanding of the dark matter/dark energy question. In this respect we note that ignoring the time dependence of $\Omega_m(y)$ and $H(y)t(y)$ is clearly misleading.

We have reached the following additional conclusions:

- If one assumes the flat Λ**CDM** model, the value of the cosmological constant is completely determined by Einstein's equation and its current value depends only on the estimations of the values of H_0 and t_0.

- The time usually associated to the formation of the CMB radiation (379,000 years after the Big Bang) was computed assuming that the Λ**CDM** model is the correct one and using a slightly smaller decoupling temperature (about 2970 K and $z = 1088.7$, which we have approximated to 3000 K and $z = 1099.7$). This time (t_{CMBR}) depends on the model used and is quite sensitive both to the universe model and to the different combinations of values of H_0 and t_0. Its value for a **KOFL** model is much larger (over one million years).

- If we start from the assumption that galaxies probably started forming after the observable universe radius exceeded the Schwartzschild radius, our tests show that, in a flat universe, many of the oldest galaxies would have been formed when the universe was still an exploding black hole. An open universe, on the other hand, gives a universe where this does not happen for $|k| > 0.9$. A mixed universe would also be compatible with this for smaller values of $|k|$ and $\Lambda_0 > 0$.

When the measurements of the apparent magnitudes of type Ia supernovae, indicative of an accelerated expansion of the universe, were reported by S. Perlmutter,[11] he signaled the possibility that part of the effect could be explained by a possible dimming by intervening dust. Later, this well-founded concern was discarded and a flat universe model has been generally assumed. However, taking into account that at early times the matter mass density was significantly higher than it is now, and so was presumably the cosmic dust density, the case for systematic corrections of the apparent magnitudes of type Ia supernovae is still worthy of consideration, together with the fact that an upwards trend at higher redshifts in the Hubble plot of $\log_{10}(r/R_0)$ versus $\log_{10}(v/c)$, as shown in Fig. A.1, could be interpreted as an accelerated expansion, completely unrelated to a non-zero cosmological constant. In this case, an open universe model could be compatible with the apparent acceleration of the universe.

Acknowledgements

We would like to thank D.N. Spergel for providing links to the complete WMAP data. We are indebted to Ralph A. Alpher and Stanley L. Jaki (no longer with us) for many fruitful conversations and protracted correspondence on cosmological matters. We are also very grateful to Manuel de la Pascua, Carmen Aragó, and Manuel I. Marqués for their support and valuable help in preparing the manuscript. And to Manuel M. Carreira S. J., José L. Sánchez Gómez, Ginés Lifante, Manuel Tello, and Antonio Alfonso Faus for many helpful conversations.

References

1. B. M. Tinsley, The cosmological constants and cosmological change, *Phys. Today* **30**, 32–38 (1977).
2. J. A. Gonzalo, J. L. Sánchez Gómez, and M. A. Alario (eds.), *Cosmología Astrofísica*, Alianza Universidad (1995).
3. A. G. Riess, W. H. Press, and R. P. Kirshner, Using type IA supernova light curve shapes to measure the Hubble constant. *Astrophys. J.* **438**, L17 (1995).
4. E. Komatsu, *et al.*, Seven-year Wilkinson Microwave Anisotropy Prove (WMAP) observations: Cosmological interpretation. *Astrophys. J. Suppl.* **192**, 18 (2011).
5. C. L. Bennett, *et al.*, Nine-year Wilkinson Microwave Anisotropy Probe (WMAP) observations: Final maps and results, (2012). arXiv:1212.5225. [astro-ph.CO].
6. NASA (2013), http://www.nasa.gov/mission_pages/planck/news/planck20 130321.html130321.html.
7. P. A. R. Ade, *et al.*, Planck 2013 results. I. Overview of products and scientific results (2013). arXiv:1303.5062. [astro-ph.CO].
8. A. G. Riess, *et al.* (Supernova Search Team), Observational evidence from supernovae for an accelerating universe and a cosmological constant, *Astron. J.* **116**(A.3), 1009–1038 (1998). arXiv:astro-ph/9805201.
9. S. Weinberg, *Cosmology* (Oxford University Press, 2008).
10. A. Einstein, *The Principle of Relativity* (Methuen, 1923; Dover Pub., 1952).
11. B. M. Schwarzschild, Discoverers of the Hubble expansion's acceleration share Nobel physics prize, *Phys. Today* **64**, 12, 14 (2011).

Appendix B

Physics and the Universe: From the Sumerians to the Late-Twentieth Century

Stanley L. Jaki

Seton Hall University, South Orange, New Jersey, USA

The monumental stillbirths which science suffered in all great ancient cultures, including the Greeks, is traced to their commonly shared pantheistic view on the universe. Within that view it was impossible to develop a science (physics) which can deal with that chief characteristic of things, which is their being in motion. The breakthrough to a viable science (physics) came in terms of a reflection on the reality of a universe created out of nothing and in time. Although in modern times this Christian notion of the universe has been largely abandoned, the best developments in science, though not in philosophies cultivated by many leading scientists, imply an endorsement of that very view.

Originally, I planned to speak on physics and the universe in the twentieth century. Professor Julio Gonzalo, the organizer of this Conference, had a different idea which, I presume, was suggested to him by my book, *Science and Creation: From Eternal Cycles to an Oscillating Universe*. In the first part of that book, first published in 1974,[a] I analyzed in seven chapters what I call the stillbirth which science suffered in all major ancient cultures, such as the Hindu, Chinese, Egyptian, Babylonian, and the Greek.

No chapter in that book was on the Sumerians. In fact, I hardly mentioned them even in connection with the Babylonians. Nevertheless, Prof. Gonzalo insisted that I begin with the Sumerians even if I talk about physics and the universe. Of course, one can gain a profound insight about the connection between the science of physics and the physical universe by looking at that connection in a historical perspective. But what about the Sumerians? The universe certainly existed at the time of the Sumerians, but physics they did not have. To give the benefit of the doubt to Professor Gonzalo, I went to the Princeton University Library and took out two books on the Sumerians. One of them seemed particularly promising as it contained a chapter on Sumerian scientists.[b] Only science, and certainly the science of physics, was missing in that chapter.

Why then do I still begin with the Sumerians? Not entirely out of courtesy for Prof. Gonzalo. I do so out of the conviction that the absence of science, and of physics in particular, among the Sumerians provides a dramatic background, a sort of indirect proof, on behalf of the thesis I am setting forth in this lecture.

In fact I can no longer say that I could care less for the Sumerians. At one point, around 1970 or so, I certainly thought that I shall always leave them alone. As I said, I mentioned them but fleetingly in my book, *Science and Creation*, a fact that caused a two-year delay, if not more, in the publication of that book. A publishing house in the USA was more than ready to take it. Indeed they were so eager as to ask a Nobel-laureate neurophysiologist, who had two years earlier given them a glowing report about my *Brain, Mind, and Computers*, to evaluate for them the manuscript of *Science and Creation*. But this time the outcome was different. The reviewer wrote a fairly long report in which he argued that I should speak at length about the Sumerians and, moreover, that I should make the Sumerians the subject of the very first chapter in my book. I demurred and had to look for another publisher. By the time the book was typeset there came the energy crisis of 1973. As a result the book was published not in 1971 but in 1974. I had reason for not being enthusiastic about the Sumerians.

Then around 1976 I attended a lecture given by the same neurophysiologist in Philadelphia. His principal thesis was the creativity of the mind, a creativity particularly evident in the arts from the earliest times. Most in the audience, including myself, had not seen before what his first slide showed: a beautiful life-sized alabaster head of a woman. My first reaction was to take it for an early Renaissance work. Amazement ran through the hall as he identified the sculpture as one made in Sumer sometime around 2400 B.C.

Here was a case of an indisputably modern view of the human head: expressive, gentle, almost sweet in spite of its dignified composure. 2000 years later Greek sculptors were just beginning to part with the standards of archaic rigidity and frozen smiles. 4500 years ago the Sumerians could display genuine modernity and in the arts where it seems to count most. The Egyptians had just started building pyramids but for a long time they did not yet sculpt or paint a female head as appealing as that Sumerian sculpture known now as the Warka Head. The Babylonians who largely inherited the Sumerian culture never came up with anything similarly close to the modern spirit.

The Babylonians continued, of course, with the two chief Sumerian inventions: the wheel (used also as potter's wheel) and cuneiform writing. The latter is a great misnomer because it makes us focus on the incidental and not on the essential. The incidental is the wedge-shaped trace left in the clay by each stroke of the iron pen used by the Babylonians. The essential is that all those wedge-shaped traces stand for letters. Now that writing has been around for 4000 years, it is difficult to appreciate what a marvel of abstraction is any letter of the alphabet. In this technological age, with instantaneous transmission and storage of images, there remains little appreciation for a far greater discovery. It is the transformation of phonetic data into abstract visual figures and their combination into groups representing words, or logoi, that are at the basis of all logic in particular and all reasoned discourse in general.

The very word, logic, should take us to the Greeks of old. They were the masters of logic in two respects: One is the development of formal logic in Aristotle's *Prior and Posterior Analytics*. The other is geometry. Euclid's fourteen books do not need our praises. Some of his propositions in the 13th and 14th books are still a challenge to first-rate mathematicians. The Greeks could be startlingly modern in other respects as well. It is enough to think of their dramatists, political theorists, and historiographers. But the Greeks, to say nothing about other great ancient cultures, failed to become modern in one very modern field, which is science.

To a somewhat less startling measure this is a fact also about the ancient Hindus, Chinese, Egyptians, and Babylonians. They all abounded in technical and speculative talent. They all were great organizers, they all interacted with other cultures, they all had material needs. But the impact of none of these factors, which are attributed decisive importance nowadays, availed them with respect to science. The Chinese of old were the inventors of the magnet, of gunpowder, and of block printing — three factors to which Francis Bacon mistakenly attributed the birth of science in early modern Europe.[c] The ancient Hindus devised the decimal system of counting, including the positional value of zero, a stunning discovery. The Egyptians of old built pyramids without pulleys and excelled in practical land surveying. Last but not least, they co-invented with the Sumerians phonetic writing.[d] The Babylonians of old compiled a vast set of astronomical observations that enabled Hipparchus to discover the precession of equinoxes.

Of course, the Greeks were most original in systematizing ideas about lines, figures, and bodies into a set of interconnected propositions. In doing so they invented the science of geometry. In more than one way geometry enabled the Greeks to find out most important things about the physical world. It should be enough to think of Eratosthenes' method of calculating the size of the earth, and of the even more ingenious method whereby Aristarchus of Samos calculated the relative distances of the earth, the moon, and the sun from one another and their absolute sizes and distances as well. Geometry worked marvels in the hands of Ptolemy whose system was as accurate in predicting the position of planets as was the system of Copernicus.[e]

While Archimedes' famed expertise in geometry served him well in working out the basic laws of hydrostatics, he is not known to have done any work in hydrodynamics. For that reason alone one may suspect that Galileo relied on much more than geometry as he specified (on the basis of a long-standing late-medieval doctrine[f]) the spaces covered in equal times by freely falling bodies or the corresponding distances on an inclined plane. Clearly, the Greeks did not possess something as they came to grips, time and again, though in vain, with the reality of motion, embodied either in the free fall of bodies or their traveling through the air as projectiles.

That something relates to the difference between two kinds of science — one that can cope with problems of statics, that is, with things on which no force acts, and the other that deals with problems of dynamics. Therein lies a principal difference between ancient times and modern times and also the clue to the failure of the Greeks of old to become truly modern. Although the same is true of other ancient cultures, in the case of the Greeks the problem can be analyzed with great precision.

Here too Aristotle provides the most instructive case. About his physics it was correctly said that it contains as many errors as pages. His errors are at times astonishing both because they are the very opposite to what Newtonian physics says and because they contradict most obvious everyday evidences. According to Aristotle the time of the fall of a body is inversely proportional to its mass. Aristotle specifically states that if two bodies are released from the same height, the one with twice of the mass of the other will reach the ground in half of the time which it takes for the other body to do so. Aristotle also states in the same context that this is so because the twice heavier body has twice as great desire to rush towards its natural place, the center of the earth.[g]

As in almost all other respects, here too, Aristotle, who loves to recall the importance of first principies, is very consistent. For him the fall of a body is but a particular case of a much vaster and more fundamental motion, or the motion of the starry heavens which in turn has its source in the very nature of the universe as he undertands it. Of course, when Aristotle says that the universe is the foremost living entity,[h] he is far from that crass animism that can be found among the Hindus, the Egyptians, and the Chinese of old, and even in the writings of some ancient Greek philosophers, especially some Neoplatonists.[i] Still the universe, as seen by Aristotle, is a subtly animated being. Its animation begins with the desire of the sphere of fixed stars for the Prime Mover, which is really not moving anything. The Prime Mover of Aristotle is not really distinct from the Empyrean Heavens, or the mythical realm above the sphere of the fixed stars.

Some interpreters of Aristotle, Thomas Aquinas, for instance, would disagree on this evaluation of the Aristotelian Prime Mover. However that may be, the desire of the heavenly sphere is, for Aristotle, a sort of divine affection because it is eternally the same. This eternal sameness is manifested in the uniform circular motion of the heavens. But as that desire is transmitted downward it gradually loses its perfection. The name, planets (or vagabonds), is an expression of this cosmic view in which a thing is the more imperfect in its motion the farther it is from the heavenly sphere. Below the moon's orb no motion can be truly regular. Consequently, Aristotle contradicts himself (as well as the facts) as he makes the rate of fall of bodies on earth a function of their mass.

In a sense Aristotle is inconsistent even when he endorses what is obvious, namely, that the fall of bodies on earth is invariably vertical. For him the reason for this is that bodies on earth share, by moving toward the center of the earth, in the perpetual trend of the universe toward self-fulfilment. But this sharing has to be very imperfect, an imperfection that rebounds on the universe itself. This can be seen when Aristotle unfolds the consequences of his pantheistic view of the universe. With his very clear idea of at least some aspects of divine perfection, Aristotle is forced to deny divinity to the world wherever it shows apparently irregular changes, which it certainly does in the sublunar or terrestrial regions. Hence the Aristotelian emphasis on the essential difference between the heavenly and terrestrial regions of the universe. The consequences of this dichotomy for physics is that there can be an exact physical science only about the heavenly regions of the universe, but not about its terrestrial parts. In other words, one can calculate exactly the motion of the heavenly sphere, and almost exactly the motion of planets, but it makes no sense to calculate the essentially irregular motion of bodies in the terrestrial realm, that is, below the moon's orbit. A dynamics, valid about the entire universe, is an impossibility, in fact a near blasphemy, as far as Aristotle's system is concerned.

It should therefore be no surprise that Aristotle says the correct things in physics only when he takes up the non-dynamical or static character of bodies, such as the direction of their free fall as necessarily perpendicular everywhere to the surface of the earth. The problem of his thoroughly mistaken dynamics should seem therefore all the more tantalizing, a word most appropriate in this connection. With his torments in Hades, Tantalus is a graphic illustration of the inner logic of the conceptual impact of the universe as conceived by the Greeks. Although eternal and therefore without beginning, that universe can have no true fulfilment. At the very moment when its heavenly movements seem to have completed a full circle (when all the planets, as Plato had suggested, are back at the same point and complete a Great Year), everything has to start all over again. The branch full of apples that hangs over Tantalus is moved by the breeze away from him just whenever he almost grabs it.

In short, the Greeks of old failed to formulate physics because their notion of the universe deprived them from making some basic conceptual breakthroughs. The matter should seem all too clear against the background of Newtonian physics which is based on Newton's three laws of motion. The first, or the law of inertial motion, was in substance formulated by John Buridan, more than three hundred years before Newton. He took his clue from the creation of the universe in time.[j] Both words, creation, and in time, are crucially important. The createdness of the universe meant for Buridan that the universe was not a necessary thing either in its existence or in its form. As a non-necessary entity, the physical universe could be for Buridan (and even more so for his successor, Oresme, at the Sorbonne) a mere clockwork in which the motion of parts was not subject to desires and other animistic factors. Such a universe could also be infinite, with no boundaries for the inertial motion of bodies, a conceptual framework very much in the mind of Descartes, whom Newton took for the first formulator of the law of inertial motion.

Both Descartes and Newton owed their notion of the physical universe to that Christian background which greatly helped Buridan and Copernicus in advancing the cause of physics. In such a universe it was possible to talk, as Newton did, in the same breath about the fall of an apple from a tree and of the fall of the moon in its orbit. In such a universe, not subject to ever variable animistic forces, it was possible to credit a form of force mathematically always identical, equal to the product of mass and acceleration, with the motion of all bodies in the universe. In such a universe, and in such a universe alone, an apple could fall on Newton's head and be a clue to the falling of bodies anywhere. Only in such a universe were great scientific geniuses free of the fate of Tantalus as they kept looking for the fruits of their research.

The story of physics since Descartes and Newton up to the present, is a story of the attitude, conscious or tacit, towards the notion, a most Christian notion, of the universe, which gave physics its viable birth in the first place. As long as nothing has been proposed or assumed, either implicitly or explicitly, by physicists against universe as a philosophical construct, the newly born physics could readily unfold its potentials. Newtonian dynamics could be extended beyond the solar system. The inverse square law was successfully applied in electricity. Heat could be treated as the motion of idealized atoms. Electricity and magnetism were united.

Such was a marvelous progress which raised physics to the highest cultural pedestal. But physics was heading toward a potential straightjacket when certain ideas incompatible with the universe were grafted on it. First came a growing scepticism about the doctrine of the createdness of the universe and of its having been created in time. An early case was Descartes himself. If not in words, at least in intention he championed the idea of a universe which was infinite, had an a priori material structure, and whose start was not so much a creative act on the part of God, but a mere flip, as Pascal already diagnosed Descartes's cosmology correctly.[k] With his a priori preference for unbounded Euclidean extension, Descartes laid the groundwork for an ever more worshipful attitude toward what later became known as the idea of an infinite Newtonian universe.[l] Such a universe began to play the role of a substitute for God from the mid-eighteenth century on for two reasons: If it was infinite, it seemed to make unnecessary that other infinite which is God.[m] And since it was infinite in a way (Euclidean three-dimensionality) which is natural for human perception, it was readily assumed to exist naturally, that is, without a Creator.[n] These were developments of great concern for theology, but here we are concerned with physics.

Physicists could have easily seen that an infinite Euclidean universe was contradictory in character. The optical paradox of an infinite Euclidean universe was openly discussed around 1700 and even in the Royal Society with Newton in the chair and with Edmund Halley as the lecturer. Halley, a well-known atheist, tried to take the sting out of the paradox and to shore up thereby the cause of the infinite universe. As it usually happens in such countertheological maneuverings, Halley too had take back with one hand from the universe what he had tried to secure to it with the other. Infinity could hardly be reconciled with the inhomogeneity which Halley, in all appearance, offered as a solution of the paradox.[o]

Just as the optical paradox was largely ignored throughout the 18th century, so was its gravitational counterpart, although already Bentley called, though only qualitatively, Newton's attention to it.[p] No physicist paid attention when in 1828 Green proposed the theory of potential, on the basis of which a strict formulation can be given to the gravitational paradox. Again, there were no notable reactions when in 1848 the younger Herschel showed that absorption of light in interstellar ether cannot solve the optical paradox. Nobody saw new opportunities for physics when in 1872 Zollner suggested in detail a four-dimensional geometry for the solution of the gravitational paradox. No physicist protested when in 1884 Lord Kelvin defended Euclidean infinity as the only form of rationality.[q] No physicist shook his head in disbelief when in 1901 Lord Kelvin proposed a solution to the optical and gravitational paradoxes on the ground that the infinite number of bright stars beyond the Milky Way was of no consequence for the visible part of the universe which he identified with the Milky Way![r]

Ultimately it was Kelvin's belief in that infinity that made him belittle as two small clouds in the bright sky of physics two problems: one posed by the black body radiation, the other by the null result of Michelson's ether-drift experiment. This worshipful respect toward the infinite Newtonian universe resulted in a schizophrenic view, subtly reminiscent of the Aristotelian cosmic dichotomy. The Newtonian infinite universe was split into two parts: one visible and puny, restricted to the Milky Way, the other the infinite beyond it and forever beyond the investigation of the physicist.[8]

For physics this meant a delay in the advent of General Relativity — for if the optical and gravitational paradoxes of an infinite universe had been kept in focus, a psychological pressure might have built up already in the eighteenth century for a consideration of the possibility of non-Euclidean geometries. It is well known that Gauss did not dare, in the early 1800's, to publish his ideas about non-Euclidean geometries lest he be taken for a fool not only by the world at large but also by the world of mathematicians and physicists. Clearly, the unjustified respect for a notion of the universe, which was wrong scientifically in the same measure in which it was anti-Christian, could but hamper the progress of physics.

No different is the twentieth-century part of the story. On the one hand we have the marvelous physics of General Relativity which assures the greatest discovery in the history of physics. The discovery is the first contradiction-free discourse about the totality of consistently interacting things or the universe. It is on that discovery that rests the rapidly expanding field of modern scientific cosmology in which one sees united the study of galaxies and stars with the study of fundamental particles. Einstein himself was rightly proud of his achievement. He saw the proof of his genius not in Special Relativity which first contained the famous formula mc^2 but in General Relativity with its consequences for cosmology.

Unfortunately for Einstein he viewed the universe as divine in a broad pantheistic sense. He therefore wanted to keep the universe essentially immobile, that is, unchanging. This is why he tried to offset its possible expansion by introducing the cosmological constant. Years later he had to admit that in doing so he committed the greatest blunder of his scientific career. He was not at all pleased when the idea and fact of the recession of galaxies had to be recognized. In order to save eternity for the universe, he then endorsed, in 1931, the idea of an oscillating universe which he had considered with great reservation nine years earlier.[t] Blinded as he was by his pantheistic respect for the universe, Einstein failed to see the obvious, namely, that the oscillations of the universe had to die out eventually.

What happened in terms of quantum theory, tells a similar story of oversight and blindness with respect to the universe. Here the first thing to note is that by 1927, when Heisenberg derived his uncertainty principle, he and a number of other physicists had explicitly rejected causality. They did so on philosophical grounds which for the most part they failed to study carefully. As is well known, around 1900 most physicists subscribed to some mixture of Humean empiricism and Kantian idealism yielding some pragmatist and operationist philosophy in which there was no room for causality in the ontological sense. In fact, some physicists were such poor philosophers at that time as to take statistical gas theory for an overthrow of causality. There is therefore much more than meets the eye in Heisenberg's statement that because of the uncertainty principle, "the principle of causality has been definitively disproved."[u]

What were the consequences for physics? I do not, of course, mean the further unfolding of the marvelous potentialities of quantum mechanics as a scientific theory and technique. In a sense, though not in exactly the same sense as Schrödinger wanted it, one can talk of quantum mechanics as "the Lord's quantum mechanics."[v] Its successes, vast and numerous, are too well known to be recalled in detail. But little has been said about the consequences for physics, for quantum mechanics in particular, of a view of the universe in which the universe is deprived of causality and becomes thereby a mere heap of random occurrences. The chief consequence for physics is that it is turned into a most dubious means whereby one can pull a rabbit from under one's hat, that is, a rabbit which was not there in the first place.

In fact not only rabbits can be "created" this way but the universe itself, and if this were not nonsense enough, several universes, indeed an uncounted number of them. The start of that hideous disrespect for reality is Heisenberg's uncertainty principle insofar as it is taken for a principle of anti-ontology and not for what it truly is, a limit put on the precision of measurements. When Heisenberg stated that the principle of uncertainty definitively disproved causality, he spoke not as a first-rate physicists but as a third-rate philosopher. He simply overlooked that what he proposed was kind of elementary fallacy for which the Greeks of old already had a technical name: *metabasis eis allo genos*. For what Heisenberg claimed, and countless physicists after him, is a jump, a metabasis, from the purely operational to the strictly ontological domain. As such the jump or move is a plain non-sequitur.

It can be rephrased as follows: An interaction that cannot be measured exactly with the tools, conceptual and instrumental, at our disposal, then such an interaction cannot take place exactly. The non-sequitur follows from taking the same word exactly, first in the operational and then in the ontological sense. Such a procedure is bordering on plain equivocation.[w]

If, however, one accepts that philosophical fallacy, one will have, in terms of Heisenberg's uncertainty principle, a pseudo-scientific proof of the foremost breakdown of causality. Such a breakdown occurs when one speaks of the creation of matter out of nothing but without a Creator. The first to graft this skullduggery on physics were the steady state theorists. They did so in support of their explicit belief in the universe as the ultimate entity, a sort of a substitute god. Like Einstein, they too wanted to save the universe from an overall change, from a possible ultimate dissolution into a practically zero density of matter. Exceedingly small was the amount they claimed to pull out of nothing into existence in order to keep the universe the same or unchangeable as befits a god. One hydrogen atom per second in every cubic mile of space is exceedingly small. But since there are an astronomically large number of cubic miles in the universe, the total amount of hydrogen thus "created" amounted to astronomically large quantities.

Revealingly, the scientific community did not attack the steady-state theorists for their misuses of philosophy. Countless physicists let themselves be pulled into the belief that the basic postulate of steady state theory has scientifically verifiable consequences, such as the extra amount of 21 cm radiation, the natural wavelength of the hydrogen atom. But even if that extra amount had been detected, it would not have proved the steady state theory. That basic postulate of the theory is beyond scientific proof or disproof. There can be no scientific proof of creation out of nothing, be that creation by God or by those parading as scientific demigods. The elementary proof of this is that the nothing is by definition unobservable even with the most refined scientific instrument.

Time does not allow me to trace here, however briefly, the development from the steady state theory to the inflationary theory and to the multiworld theory of the universe. In both, ability is claimed by physics to create universes literally out of nothing. Nor is there time here to review the anti-theistic inspiration of the inflationary theory of the universe in which the universe, this very object that gives meaning to physics, is being fragmented into an immensely large number of "universes," all disconnected from one another by the statistically different sets of laws governing each.

Here only a few words ought to be said about the impact on physics of such unconscionable games with the universe. The impact is a decrease of awareness of the possible shortcomings of quantum theory. For if quantum mechanics is taken by the physicist for an assurance that, to quote a professor at MIT, he can create universes literally out of nothing,[x] then he will hardly pay attention to the warning which Dirac gave at the Jerusalem Einstein conference in 1979[y]:

> It seems clear that present quantum mechanics is not in its final form ... Some day a new relativistic quantum mechanics will have determinism in the way that Einstein wanted. This determinism will be introduced only at the expense of abandoning some other preconceptions which physicists now hold, and which it is not sensible to try to get at now ... So under these conditions I think it is very likely, or at any rate quite possible, that in the long run Einstein will turn out to be correct even though for the time being physicists have to accept the Bohr probability interpretation — especially if they have examinations in front of them.[y]

I did not quote Dirac in order to suggest that it is the fault of quantum mechanics that it cannot cope with General Relativity. It may be the other way around. Quantum mechanics, as it stands, is a wonderful physics, though limited by its very basic assumption, namely, that atomic processes can only be dealt with in groups, and not invididually, and only in terms of a statistical theory steeped in non-commutative algebra. Indeed, as Aspects's experiments showed, Bell's theorem is correct in predicting that on the atomic level quantum mechanical statistics fits the phenomena better than does classical statistics. This is, however, merely a proof that quantum mechanics, as it stands, cannot be complemented with a theory of hidden variables, but not a disproof of the theoretical possibility of a new form of quantum mechanics as suggested by Dirac.

But neither quantum mechanics, nor any form of physics, however successful, can assure the physicist that he possesses the last word about the physical universe. As an experimentalist, the physicist can never be sure that the universe has no more hidden aspects to it. As a theoretician, the physicist should be mindful of Godel's incompleteness theorems which forclose the derivation of the consistency of an all-inclusive physical theory on a priori grounds.[z] Yet the physicist must assume the existence of a universe, fully consistent in its operations and fully investigable scientifically, if the science of physics has to retain its meaning.

About that universe no physics, Newtonian or modern, provided deeper insight than the one offered by theology. If we turn to John Henry Newman in this centenary of his death for such an insight, we will not be disappointed. In his *The Idea of a University* he wrote: "There is but one thought greater than that of the universe, and that is the thought of its Maker."[aa] And if we turn to that theologian, Thomas Aquinas, whose timeliness depends even less on anniversaries, we can hear the same praise of the universe in terms even more startling: "God himself could not have created something greater than the universe."[ab] Only such an exalted view of the universe can keep physics in a safely exalted position.

Endnotes

a A second enlarged edition was published, also by Scottish Academic Press, Edinburgh, in 1986 and reissued in the USA by the University Press of America, Lanham, Md., in 1990.

b i. E. Lansing, *The Sumerians: Inventors and Builders* (McGraw-Hill, New York, 1978).

ii. L. Cottrell, *The Quest for Sumer* (G. P. Putnam's Sons, New York, 1965), the handling of Sumerian "science" consists in references to the "science" of Babylonians of Hammurabi's time. Both works heavily depend on.

iii. S. N. Kramer *History Begins at Sumer* (1959).

c S. L. Jaki, *The Origin of Science and the Science of Its Origin* (Scottish Academic Press, Edinburgh, 1978), pp. 7–16, on the thoroughly mistaken ideas of Bacon on the origin of science.

d This point is made because of the widespread custom of taking Egyptian hieroglyphics for a form of ideographic writing, similar to the one still used in China.

e The reference is, of course, to the system of Copernicus himself and not to the subsequently much improved forms of heliocentric planetary theories.

f The doctrine in question is known as the impetus theory, first proposed by John Buridan at the Sorbonne around 1330 and discovered by Pierre Duhem around 1906 through his pioneering study of medieval manuscripts. By the time Copernicus relied on the impetus theory to solve the physical problems arising from the twofold motion of the earth, the impetus theory had been so generally accepted as to make unnecessary references to its original formulator.

g See *On the Heavens*, pp. 273b–274b.

h See *On the Heavens*, p. 285a.

i A particularly glaring example is *De antro nympharum* of Porphyry (233–C.304) disciple and editor of Plotinus.

j S. L. Jaki, *The Savior of Science* (Regnery Gateway, Washington D. C., 1988), pp. 52–53.

k "I cannot forgive Descartes. In all his philosophy he would have been quite willing to dispense with God. But he had to make Him give a fillip to set the world in motion; beyond this, he has no further need of God." B. Pascal, *Pascal's Pensees,* translated by W. F. Trotter, with an Introduction by T. S. Eliot (E. P. Dutton, New York, 1958), p. 23.

l The idea was un-Newtonian in that Newton held the physical universe to be finite, surrounded by a no less finite layer of the ether, beyond which was the infinite, though non-physical, space, taken by Newton for the "sensorium of God."

m Such a reasoning, cheerfully accepted by agnostics in the nineteenth century, still caused perplexities to eighteenth-century deists, as is clear from the article "Infini" in the *Encyclopedia* of Diderot and d'Alembert, where it is clearly stated that only God can be strictly infinite.

n The same reasoning reappeared in our time in the cosmology of steady-state theorists and, more memorably, in the concluding chapter of S. Hawking, *A Brief History of Time.* See S. L. Jaki, Review of Hawking's book, "Evicting the Creator," in *Reflections* (Spring 1988), reprinted in S. L. Jaki, *The Only Chaos and Other Essays* (University Press of America, Lanham, Md.; Intercollegiate Studies Institute, Bryn Mawr, PA., 1990), pp. 152–161.

o For details, see S. L. Jaki, *The Paradox of Olbers' Paradox* (Herder and Herder, New York, 1969), Chapter 4.

p S. L. Jaki, *The Paradox of Olbers' Paradox,* pp. 60–62.

q To make matters worse, Lord Kelvin did so in the context of a much publicized lecture in Philadelphia. See S. L. Jaki, *The Paradox of Olbers' Paradox,* p. 170.

r For details, see S. L. Jaki, *The Milky Way: An Elusive Road for Science* (Science History Publications, New York, 1972), pp. 276–277.

s The topic of Chapter 8, "The Myth of One Island," in S. L. Jaki, *The Milky Way.*

t For details, see S. L. Jaki, *Science and Creation,* p. 357.

u The phrase is from Heisenberg's paper of 1927 containing the formula for the principle of uncertainty.

v E. Schrodinger, *What Is Life and Other Scientific Essays* (Doubleday, Garden City, New York, 1956), p. 83.

w A point on which I have repeatedly insisted for now over a dozen years, most recently in S. L. Jaki, Determinism and reality, *Great Ideas Today 1990* (Encyclopedia Britannica, Chicago, 1990), pp. 278–302.

x The professor in question is A. H. Guth, who made this preposterous claim both in published interviews and in scientific publications. For details, see S. L. Jaki, *God and the Cosmologists* (Regnery Gateway, Washington, D.C., 1989), p. 258.

y Dirac statement was quoted by R. Resnick, a participant in that Conference, in his article, Misconceptions about Einstein: His work and his views," *Journal of Chemical Education* **52** (1980) 860.

z For a detailed discussion, see Chapter 4 "Godel's shadow" in S. L. Jaki, *God and the Cosmologists.*

aa See the 8th edition of S. L. Jaki, *My God and the Cosmologists* (Longmans, Green and Company, London), p. 462.

ab For details, see S. L. Jaki, Thomas and the Universe, in *The Thomist* **53** (1989) 545–572.

"Cosmos & Chaos" (Rome, 2019)

Julio A. Gonzalo[*] and Manuel Alfonseca[†]

Universidad Autónoma/AECyC, Madrid

[†] *Lucía Guerra Menéndez, Universidad San Pablo CEU, Madrid*

Exact quantitative analytical solutions of Einstein's cosmological equations for a **finite open** (KOFL) Universe and a **finite flat** (LCDM) Universe, result in respectively in total cosmic masses $M \simeq 1.7 \times 10^{54}$g($k < 0$, $\Lambda = 0$) and $M \simeq 9.5 \times 10^{55}$g very large but **finite**. The first value is obtained with $H_0 = \dot{R}_0/R_0$ as given by 67.74 ± 0.46 kms^{-1}Mpc^{-1} (PLANCK, 2013) recently confirmed by 69.0 ± 0.5 kms^{-1}Mpc^{-1} (Gravitational Waves, 2017). This results in $H_0 t_0 < 1$ (dimensionless). On the other hand, a different value is obtained using $H_0 = 73.24 \pm 1.74$ (Supernovae 2016) resulting in $H_0 t_0 > 1$ which excludes the possibility of an open $k < 0$, $\Lambda = 0$ Universe. In both cases $t_0 = 13.7 \pm 0.2$ Gyrs (WMAP 2003/PLANCK 2013) should be used giving the time elapsed between the Big Bang and today. According to Fr. Jaki (in full agreement with Einstein and Lemaitre, and other prominent cosmologists), the Universe must necessarily be **finite** on the grounds of **logical** and **metaphysical** principles. Alan Guth and others have proposed a flat Inflationary Universe which might be infinite. Most theoretical cosmologists today support a flat LCDM Universe which does not exclude the possibility of an infinite self-sufficient Lambda Cold Dark Matter Universe or Multiverses.

1. Introduction

As noted recently by Julio A. Gonzalo and Manuel Alfonseca most theoretical cosmologists give preference to the LCDM (Lambda Cold Dark Matter) model based substantially in the Inflationary hypothesis of Alan Guth.[1] This model assumes zero space-time curvature $(k = 0)$ and a non-zero cosmological constant $(\Lambda > 0)$ resulting in cosmic accelerated expansion. In addition to the cosmic density parameter $\Omega_0 = \rho_m(t_0)/\rho_{cm}(t_0)$, where $\rho_{cm}(t_0) = 3H_0/8\pi G$ is the critical mass density parameter (dimensionless), another important dimensionless cosmic parameter is the product $H_0 t_0$ dependent on the Hubble–Lemaitre ratio $H_0 = \dot{R}_0/R_0$. As has been noted previously[2] the value of H_0 determined from the CBR (Cosmic Background Radiation) is significantly different from that determined from Supernovae data.[2] The value of t_0, time elapsed since the Big Bang to today is very accurately known from NASA'S WMAP satellite data and ESA'S PLANCK satellite data[3] which coincide within less than 1%. On the other hand, very recent data on H_0 extracted from the collapse of two distant very massive objects, agrees very well with the CBR value determined by PLANCK'S data, but not with the Supernovae value, as shown below in Table C.1.

It is very important to note that the compact rigorous[4] solutions of Einstein's Cosmological Equations for finite KOFL **open** Universe $(k < 0, \Lambda = 0)$ lead to

$$H(y)t(y) = \frac{1}{\tanh^2 y} - \frac{y/\tanh y}{\sinh^2 y} \rightarrow \frac{2}{3} \leq H(y)t(y) \leq 1 \qquad \text{(C.1)}$$

for y going from $y = 0$ (Big Bang) to $y \rightarrow \infty$ (very distant future). On the other hand, data for a finite **flat** Universe $(k = 0, \Lambda > 0)$ of Einstein's Cosmological Equations lead to

$$H(y)t(y) = \frac{2}{3}\frac{y}{\tanh y} \rightarrow \frac{2}{3} \leq H(y)t(y) \rightarrow \infty \qquad \text{(C.2)}$$

Table C.1.   Recent data on H_0 extracted from the collapse of two distant very massive objects.

H_{01}(Supernovae)	$= 73.24 \pm 1.47\,\text{kms}^{-1}\text{Mpc}^{-1}$	$\rightarrow H_0 t_0 = 1.026 \pm 0.240$
H_{02}(CBR)	$= 67.24 \pm 0.46\,\text{kms}^{-1}\text{Mpc}^{-1}$	$\rightarrow H_0 t_0 = 0.950 \pm 0.016$
H_{03}(Grav. Waves)	$= 69.00 \pm 0.50\,\text{kms}^{-1}\text{Mpc}^{-1}$	$\rightarrow H_0 t_0 = 0.966 \pm 0.016$

for y going from $y = 0$ (Big Bang) to $y \to \infty$ (very distant future) resulting in $H(y)t(y) \to \infty$.

So $H(y_0)t(y_0)$ is allowed to be ≤ 1 for an **Open** Universe but $H(y_0)t(y_0) > 1$ would imply necessarily a **Flat** Universe. This appears to be precluded by the very recent H_{03} (Gravitational Waves) value, in such a good agreement with the H_{02} (CBR) value.

It must be noted that the compact rigorous solutions of Einstein's Cosmological Equations for a **flat** Universe lead to a **finite mass** for the Universe only slightly larger than the mass of the **open** Universe, evidently **finite** as assumed originally by Einstein and Lemaitre, among other prominent cosmologists.

In the next section we will examine quantitatively the case for a KOFL **Open Universe** ($k < 0$, $\Lambda = 0$) and for a LCDM **Flat Universe** ($k = 0$, $\Lambda > 0$), and we will make finally the pertinent concluding remarks, having into account Fr. Jaki's view on the subject.

2. The Case of an Open Universe

Einstein's cosmological Equations can be summarized[3] by

$$\dot{R}^2 = \frac{8\pi}{3}G\rho R^2 - kc^2 + \frac{\Lambda}{3}c^2 R^2 \tag{C.3}$$

where R is the cosmic radius, $\dot{R}$ its time derivative, G Newton's gravitational constant ($G = 6.67 \times 10^{-8}$ in cgs units), ρ the mass density (including matter mass and radiation mass), k the space-time curvature, c the speed of light in vacuum ($c = 3 \times 10^{10}$ cm/s), and Λ the so called Einstein's cosmological constant.

In principle k could be $k < 0$, $k = 0$, and $k > 0$. The case of $k > 0$ (closed universe) may be excluded from further consideration because it requires that $H(y_0)t(y_0) < \frac{2}{3}$, contrary to the observational evidence.

For an open ($k < 0$) Universe assuming $\Lambda = 0$, something which Einstein did regret not to have done after discussing the matter with Lemaitre & Hubble and accepting the available observational evidence Eq. (C.3) becomes

$$\dot{R}^2 = \frac{2GM}{R} + |k|c^2. \tag{C.4}$$

We can define

$$\frac{2GM}{R_+} = |k|c^2, \quad R_+ = \frac{2GM}{|k|c^2} \tag{C.5}$$

and can integrate the nonlinear differential equation[3] arriving to

$$t(y) = \frac{R_+}{c|k|^{1/2}} [\sinh y \cosh y - y] \tag{C.6}$$

$$R(y) = R_+ \sinh^2 y \tag{C.7}$$

From Eqs. (C.6) and (C.7) we can get directly, having into account that $H(y) = \left[\frac{\mathrm{d}}{\mathrm{d}y} R(y) \middle/ \frac{\mathrm{d}}{\mathrm{d}y} t(y) \right] \middle/ R(y)$, the dimensionless product

$$H(y)t(y) = (\tanh y)^{-2} - (y/\tanh y)(\sinh y)^{-2} \tag{C.8}$$

which allows us to get y_0 knowing[5] $H(y_0) = 69.0 \pm 0.5 \, \mathrm{kms}^{-1}\mathrm{Mpc}^{-1} = 2.995 \times 10^{-18} \, \mathrm{s}^{-1}$, and $t_0 = 13.7 \pm 0.2 \, \mathrm{Gyrs} = 4.32 \times 10^{17} \, \mathrm{s}$, resulting in

$$y_0 = 2.335 \tag{C.9}$$

Then we get $R_+/c|k|^{1/2}$ from Eq. (C.6) knowing, as we know, t_0 resulting in

$$R_+/c|k|^{1/2} = t_0[\sinh y_0 \cosh y_0 - y_0]^{-1} = 1.700 \times 10^{16}\mathrm{s} \tag{C.10}$$

Having into account that

$$\dot{R}(y_+) = c|k|^{1/2}/\tanh y_+ = c \tag{C.11}$$

and that at $R = R_+$ we have $\sinh(y_+) = (R/R_+)^{1/2} = 1$ and we get

$$y_+ = \sinh^{-1}(1) = 0.8813$$

and therefore that

$$|k|^{1/2} = \tanh y_+ = 0.70706 \quad \rightarrow \quad |k| = \frac{1}{2} \qquad \text{(C.12)}$$

This implies that

$$R_+ = c \left(\frac{1}{2}\right)^{1/2} t_0 [\sinh y_0 \cosh y_0 - y_0]^{-1} = 3.606 \times 10^{26} \text{ cm}$$

$$\text{(C.13)}$$

and

$$R_0 = R_+ \sinh^2 y_0 = 9.832 \times 10^{27} \text{ cm} \qquad \text{(C.14)}$$

The density parameter $\Omega = \rho_m/\rho_{cm}$, where $\rho_m = M / \frac{4\pi}{3} R^3$ and $\rho_{cm} = 3H^2/8\pi G$, comes out to be $\Omega = 1/\cosh^2 y$, the redshift $z = (R_0/R) - 1$, and the cosmic background temperature $T = R_0 T_0 / R$ after decupling, which is fixed[6] very accurately using NASA'S COBE satellite data $T_0 = 2.726\,°K$.

Table C.2 below gives the evolution of cosmic quantities as we go back in time from the present $(z = 0)$ to protogalaxy formation[7] $(z \simeq 10)$ to the time at which $\dot{R}_+/c = 1$, corresponding to $\Omega_+ = \frac{1}{2}$ $(z_+ = 26.2)$, to decoupling (atom formation) time: $\rho_m = \rho_r$.

Knowing R_+ and $|k|$ it is possible to determine the total mass of the open universe $(\Lambda = 0)$ as

$$M = |k|c^2 R_+/2G = \left(\frac{4\pi}{3} R_0^3\right) \Omega_0 (3H_0^2/8\pi G) = 1.717 \times 10^{54} \text{ g}$$

$$\text{(C.15)}$$

which, of course, is very large but **finite**. This value is reasonable, since we know that there are about 10^{11} galaxies each with about 10^{11} stars in the universe. The average mass of a typical star would be then about 10^{32} g, not far from the mass of the sun, $M_\ominus \simeq 2 \times 10^{32}$ g.

Table C.2. Evolution of cosmic quantities as we go back in time from the present $(z = 0)$ to protogalaxy formation $(z \simeq 10)$ to the time at which $\dot{R}_+/c = 1$, corresponding to $\Omega_+ = \frac{1}{2}$.

y	t(s)	H(s^{-1})	Ht	Ω	R (cm)	$\dot{R}/c$	z	T(°K)
2.355	4.32×10^{17}	2.197×10^{-18}	0.949	0.035	0.983×10^{28}	0.7199	0	2.726
1.235	0.1805×10^{17}	29.59×10^{-18}	0.851	0.287	0.0632×10^{28}	0.8377	10.0	42.0
0.8813	0.0958×10^{17}	83.20×10^{-18}	0.753	0.500	0.0360×10^{28}	1.0000	26.2	74.2
0.2044	0.0345×10^{17}	7025×10^{-18}	0.685	0.959	0.00152×10^{28}	3.5074	651	1778

3. The Case of a Flat Universe with $\Lambda > 0$

In this case Einstein's Cosmological Equations reduce to

$$\dot{R}^2 = \frac{8\pi}{3}G\rho R^2 + \frac{\Lambda}{3}c^2 R^2 \qquad (C.16)$$

which has compact solutions[2] given by

$$t(y) = \frac{2}{3}\left(\frac{\Lambda}{3}c^2\right)^{-1/2} y \qquad (C.17)$$

$$R(y) = R_L \sinh^{2/3} y \qquad (C.18)$$

From Eqs. (C.17) and (C.18) we get the pertinent dimensionless product

$$H(y)t(y) = \frac{2}{3}\frac{y}{\tanh y} \qquad (C.19)$$

and using now $H(y_0) = 73.24 \pm 1.74\,\mathrm{kms}^{-1}Mpc^{-1}$, and again $t_0 = 13.7 \pm 0.2\,\mathrm{Gyrs} = 4.31 \times 10^{17}$ s, we get

$$y_0 = 1.350 \qquad (C.20)$$

Then we can get $\frac{2}{3}\left(\frac{\Lambda}{3}c^2\right)^{-1/2}$ from Eq. (C.17) for $t = t_0$ resulting in

$$\left(\frac{\Lambda}{3}c^2\right)^{-1/2} = \frac{3}{2}t_0/y_0 = \frac{3}{2}(4.32 \times 10^{17})/(1.35) = 3.2 \times 10^{17} \qquad (C.21)$$

which implies $\Lambda = 5.89 \times 10^{-30}$, and taking into account that for $y_{Sch} = 0.8813$

$$\dot{R}(y) = \frac{dR/dy}{dt/dy} = \frac{R_L}{\left(\frac{\Lambda}{3}c^2\right)^{-1/2}}\frac{\sinh^{2/3} y}{\tanh y} = \frac{R_L}{\left(\frac{\Lambda}{3}c^2\right)^{-1/2}}\frac{\sinh^{2/3} y_{Sch}}{\tanh y_{Sch}} = c \qquad (C.22)$$

Table C.3. Evolution of cosmic quantities going back from present ($z = 0$) to $z_{\max} = 0.5341$.

y	$t(s)$	$H(s^{-1})$	Ht	Ω	R (cm)	$\dot{R}/c$	z	$T(K)$
1.35	4.32×10^{17}	2.375×10^{-18}	1.026	0.236	1.420×10^{28}	0.7948	0	2.726
0.8813	2.82×10^{17}	3.243×10^{-18}	0.9145	0.5000	0.925×10^{28}	1.000	0.534	4.184
1.08×10^{-3}	3.45×10^{13}	1.929×10^{-18}	0.666	0.9999	0.0021×10^{28}	20.99	675	1842

where $\sinh y_{Sch} = 1$, we finally get

$$R_L = R_L \sinh^{2/3}(y_{Sch}) = \left(\frac{\Lambda}{3}c^2\right)^{-1/2} c = 9.26 \times 10^{27} \text{ cm} \qquad (C.23)$$

The present radius of this flat universe is then

$$R_0 = R_L \sinh^{2/3} y_0 = 1.420 \times 10^{28} \text{ cm} \qquad (C.24)$$

The density parameter is again $\Omega = 1/\cosh^2 y$, the redshift is $z = (R_0/R) - 1$, and the cosmic background temperature is again given by $T = T_0 R_0/R$, after decoupling, with $T_0 = 2.726\,°$K.

Table C.3 gives the evolution of cosmic quantities going back from the present ($z = 0$), but, since $z_{Sch} = 0.5341$ is **less** than $z \simeq 10$ for protogalaxy formation, no room for galaxy formation is allowed in a LCDM flat Universe according to the previous analysis. Decoupling (atom formation) takes place in a flat universe at a temperature close to $2000\,°$K.

The total mass of this flat universe ($k = 0$, $\Lambda > 0$) can be determined from

$$M = \left(\frac{4\pi}{3}R_0^3\right)\Omega_0\left(\frac{3H_0^2}{8\pi G}\right) = 0.948 \times 10^{55}\,\text{g} \qquad (C.25)$$

which is not very dissimilar to M for an **open** universe as given by Eq. (C.14), and is very large, but **finite**.

4. Concluding Remarks

Finally let us say that the most accurate observational values of t_0 and H_0 as given by the **CBR** and **Gravitational Waves** favour finite **Open Universe** with $k = 0$, $\Lambda = 5.89 \times 10^{-30}$ which implies a maximum redshift $z_m = 0.534 < z_{pg} \simeq 10$.

We can conclude with an extensive quotation of Fr. Jaki at the end of his "**Postcript**" to the 2nd ed. of "God and the cosmologists"[8]:

> *"It therefore remains largely a matter of intellectual courage to stand up for the validity of the ontological sense of the question, "why such and not something else?" as it is posed by any **finite** (underlined by JAG) thing, be that thing the Universe itself. It takes even greater courage, although it should seem a mere matter of logic, to vindicate the mind's rights to a truly satisfactory answer posed by any finite existent, which the Universe certainly is. The Universe if finite at least in the sense of being restricted to a very narrow set of parameters. Not all that is conceivable does exist. Curiosity about this fact in the ontological sense is what evokes God, the Ultimate being in intelligibility, anywhere but especially within the framework of cosmology or the study of the Universe, which is supposed to be the All, a coherent Totality. Devotees of incoherence have not legitimate place in science, let alone in the science of cosmology as long as — **logy** is tied to **logic**, and the latter to **logos**, and cosmos stands for a coherent **all** and not for a scientific fad. The **All** is the Universe, write large. Not being necessarily what it is, such Universe remains a stubbornly vivid pointer to God. He is the only explanation why the All is not a glorified chaos but a cosmos which cosmologists, though they cannot create anything by any stretch of imagination, are specially privileged to investigate".*
> March 1998 S.L.I.

We have seen in this work, presented here to commemorate the 10th anniversary of Fr. Jaki's decease that, from a rigorous astrophysical perspective, that the data obtained by first class contemporary experimentalists measuring gravitons ejected by a distant colliding pair of neutron stars support a value for the Hubble–Lemaitre parameter consistent with a finite Universe, in agreement with no others than Einstein and Lemaitre, as well as with Fr. Jaki.

References

1. Alan Guth, *The Inflationary Universe* (Perseus Books: Cambridge Massachusetts, 1977).
2. See Appendix A, Julio A. Gonzalo, *Cosmic Paradoxes*, 2nd Ed. (World Scientific, Singapore 2017) and references therein.

3. P. A. R. Ade *et al.*, PLANCK 2013 results, I. Overview of products and scientific results (2013). arXiv: 1303.5062 (astro-ph-Co)
4. Julio A. Gonzalo, *Cosmic Paradoxes*, Chap. 11, pp. 69–76.
5. D. Holz, S. Hughes and B. Schutz, *Physics Today* **71**, 12 (2018), pp. 34–40 and references therein.
6. Julio A. Gonzalo, *The Intelligible Universe: An Overview of the Last Thirteen Billion Years* (World Scientific, Singapore, 2008) and references therein.
7. Joseph Silk, *The Big Bang*, 3rd. Ed. (W. H. Freeman and Company: New York, 2001) and references therein.
8. S. L. Jaki, *God and the Cosmologists* (Real View Books: P. O. Box 1793, Fraser, Michigan, 48048) p. 271.

Werner Heisenberg (1901–1976)

Georges Lemaitre (1894–1966)

Appendix D

On the Heisenberg-Lemaitre time vs. Planck's time

Let us consider[1] a massive particle with a mass given by m_{Pl}, and a radius l_{Pl} such that its gravitational self-energy equals its relativistic energy as given by Einstein's relation:

$$m_{Pl}c^2 = \frac{Gm_{Pl}^2}{l_{Pl}}, \tag{D.1}$$

Heisenberg's uncertainty principle (assuming $\Delta p \cong p$ and $\Delta x \cong x$) applied to this particle leads to

$$m_{Pl}c \cdot l_{Pl} \cong \hbar \tag{D.2}$$

and combining Eqs. (D.1) and (D.2) we get

$$(m_{Pl}c \cdot l_{Pl}) \cdot c \cong \hbar c \cong Gm_{Pl}^2, \tag{D.3}$$

i.e.

$$m_{Pl} \cong (\hbar c/G)^{\frac{1}{2}} \cong 2.17 \times 10^{-5}\,\text{g} \tag{D.4}$$

Planck's mass is therefore much larger than a baryon (proton, neutron) mass, $m_b \cong 1.67 \times 10^{-24}$ g.

Other physical quantities of Planck's particle can be obtained directly

$$l_{Pl} \cong \hbar/m_{Pl}c \cong (\hbar G/c^3)^{1/2} = 1.61 \times 10^{-33}\,\text{cm} \tag{D.5}$$

$$t_{Pl} \cong l_{Pl}/c \cong (\hbar G/c^5)^{1/2} = 5.36 \times 10^{-44}\,\text{s} \tag{D.6}$$

$$T_{Pl} \cong m_{Pl}c^2/2.8k_B = (\hbar c^3/G)^{1/2}/2.8k_B$$
$$= 5.05 \times 10^{31}\,\text{K} \tag{D.7}$$

Table D.1.   Planck's "natural unit".

Cosmic Quantity	Planck Natural Unit	Ratio
$M_u \cong 0,70 \times 10^{55}$ g	$m_{Pl} = 2.17 \times 10^{-5}$ g	3.22×10^{59}
$R_+ \cong 5.31 \times 10^{26}$ cm	$l_{Pl} = 1.61 \times 10^{-33}$ cm	3.29×10^{59}
$t_+ = 1.88 \times 10^{16}$ s	$t_{Pl} = 5.36 \times 10^{-44}$ s	3.52×10^{59}

All these quantities are given in terms of the fundamental physical constants $\hbar$ (Planck quantum of action), G (Newton's gravitational constant), c (velocity of light in vacuum), and k_B (Boltzmann's constant). Planck's called them "natural units" and we can check that for our universe, which for Einstein and Lemaitre was **finite** (and could not be otherwise) they give surprisingly concordant numbers as shown in Table D.1. It will be seen in Chapter 11 that the presently observed Hubble's ratio (time dependent) $H_0 \cong 69$ Km/s/Mpc and the present time (the time elapsed since the Big Bang to today) $t_0 = 13.8 \times 10^9$ years, allows one to use the solutions of Einstein's cosmological equations to get numbers for the total mass of the universe and for its characteristic (Schwarzchild) radius, and characteristic time

$$M_u = 0.7 \times 10^{55} \text{ g}, \quad R_+(t_+) = GM/c^2 = 5.3 \times 10^{26} \text{ cm}$$

$$t_+ = 600 \,\text{Myrs} = 1.88 \times 10^{16} \text{ s}$$

The present radius and present time are $R_0 \cong 1.5 \times 10^{28}$ cm, $t_0 \cong 13.8$ Gyrs

Alternatively, taking the total cosmic mass M_u as the starting point, we can construct a set of units, which we will call Heisenberg–Lemaitre units, as follows

$$M_u = c^2 R_+/G = 0.70 \times 10^{55} \text{ g} \tag{D.8}$$

$$l_{HL} = c \cdot (Mc^2/\hbar) = 1.40 \times 10^{-112} \text{ cm} \tag{D.9}$$

$$t_{HL} = M_u c^2/\hbar = 0.46 \times 10^{-102} \text{ s} \tag{D.10}$$

$$T_{HL} = M_u c^2/2.8 k_B = 1.63 \times 10^{91} \text{ K} \tag{D.11}$$

Equation (D.10) sets a minimum time beyond which it becomes meaningless to speculate about cosmic dynamics. At lower times, the

uncertainty principle

$$\Delta M_u c^2 \cdot \Delta t_{HL} > \hbar \qquad (D.12)$$

forbids further speculation. In a sense, this takes care of the singularity at $R = 0$, $t = 0$. Heisenberg–Lemaitre's time $t_{HL} = 0.46 \times 10^{-102}$ s is many orders of magnitude smaller than Planck's time $t_{Pl} = 5.36 \times 10^{-44}$ s.

It may be noted[2] that the solution of Einstein's cosmological equation for an expanding universe implies at very early time that cosmic radius $R(t)$ grows with time as

$$R(t) = \text{Const.} \times t^{2/3} (\text{const} = R_+ [|k|^{1/2} c/R_+]^{2/3} \qquad (D.13)$$

This means that, according to Einstein's compact solutions for an expanding universe, the growth factor between $t = t_{HL} = 0.46 \times 10^{-102}$ s and $t = t_{\text{infl}} \cong 10^{-35}$ s, as given by Alan Guth,[3] is of the order of

$$R(t_{\text{infl}})/R(t_{HL}) \cong (t_{\text{infl}}/t_{HL})^{2/3} = 7.8 \times 10^{44} \qquad (D.14)$$

which is of the order of the inflationary growth factor ($\approx 10^{40}$) assumed in Inflationary Cosmology to take place almost instantaneously at $t = t_{\text{infl}}$ in a singular cosmic phase transition. Equation (D.13) gives a growth factor of the same order taking place continuously between 10^{-102} and 10^{-35} s which, it should be admitted, is not very different from "instantaneously". The growth takes place smoothly according to the compact solution of Einstein's cosmological equation. Therefore, no cosmic first order phase transition is needed.

Table D.2 compares the Compton radius ($r_c = \hbar/mc$) and the Schwarzschild radius ($r_{Sch} = Gm/c^2$) for massive objects from the whole universe to the almost massless neutrino.

For a mass $m_{Pl} = (\hbar c/G)^{1/2} = 2.17 \times 10^{-5}$ g, the Compton radius $r_{Compton} = \hbar/m_{Pl}c = 1.61 \times 10^{-33}$ cm is equal to its Schwarzschild radius $r_{sch} = Gm/c^2 = 1.61 \times 10^{-33}$ cm.

In the 1950's the **"Steady State Theory"**[4] of Gold, Bondi and Hoyle postulated "continuous creation of energy out of nothing" in order to keep constant cosmic density throughout the expansion for a universe in continuous growth. For years, the **"Steady State**

Table D.2. Compton and Schwarzschild radii for massive objects.

Object	$m(g)$	$r_{Compton}(cm)$ $\hbar/mc$	$r_{Sch}(cm)$ (Gm/c^2)
Universe	0.70×10^{55}	5×10^{-93}	5.2×10^{26}
Galaxy	10^{44}	3.5×10^{-82}	7.4×10^{15}
Star	10^{33}	3.5×10^{-71}	7.4×10^{4}
Earth	6×10^{24}	5.8×10^{-63}	4.4×10^{-4}
Planck unit	2.1×10^{-5}	$\mathbf{1.61 \times 10^{-33}}$	$\mathbf{1.61 \times 10^{-33}}$
Baryon	1.67×10^{-24}	2.1×10^{-14}	1.2×10^{-52}
Electron	9.1×10^{-28}	3.8×10^{-11}	6.7×10^{-56}
Neutrino	2.2×10^{-35}	1.5×10^{-3}	1.6×10^{-63}

Theory" was a rival of the **"Big Bang Theory"** originated in the **"Primeval Atom Theory"** of Lemaitre and then reformulated by Gamov, Alpher and Herman.[5] The discovery of the Cosmic Background Radiation[6] by Penzias and Wilson dealt a serious blow to the "Steady State Theory" which was unable to handle that cosmic radiation.

In 1993 at the time of the El Escorial Summer Course on Astrophysical Cosmology,[7] t_0, the age of the universe (time elapsed since the Big Bang) was located somewhere between ten and twenty billion years and the numerical value of Hubble's ratio $H_0 = \dot{R}_0/R_0$ somewhere between 50 and $100 \, \text{km/s/Mpc}$. The interaction there with Ralph Alpher, John Mather, George Smoot, Stanley Jaki and others at the conference and after was extremely useful to accurately anticipate both quantities as $t_0 = 13.7 \times 10^9 \, \text{yrs}$ and $H_0 = 65 \, \text{km/s/Mpc}$, in "Acta Cosmologica" (Cracow) in 1998, and it was reprinted five years later as an Appendix[8] in "Inflationary Cosmology Revisited".

At El Escorial, I asked George Smoot (after his talk on "COBE's observations of the Early Universe") what did he think about the formal equivalence of the equations in the **Inflationary Theory** describing **sudden** cosmic expansion at constant density with the equations in the old **"Steady State Theory"** describing **continuous** cosmic expansion at constant density. I pointed out that it looked to me as if in the "Inflationary Theory" the same non energy conserving process was simply moved back in time, well beyond the reach of any possible observation. He answered, if I remember correctly, that in his opinion the two processes were not exactly identical, without denying certain similarities.

In year 1927, Werner Heisenberg published[9] his seminal work on the uncertainty principle. It was also the year in which Georges Lemaitre published[10] his original work on the solutions of Einstein's cosmological equation for a finite universe which opened the way for the Big Bang model.

Thermal History of the Universe

We know that in an **open, finite** ($k < 0$, $\Lambda = 0$) universe, at very early times, the solutions of Einstein Cosmological Equations simplify in such a way that with the dimensionless cosmological parameter $y \ll 1$ they become

$$R(y) = R_+ \sinh^2 y \xrightarrow[y \ll 1]{} R_+ y^2 \qquad (D.15)$$

where $R_+ = 4.58 \times 10^{26}$ cm, and

$$t(y) = [R_+/c\,|k|^{1/2}][\sinh y \cosh y - y] \xrightarrow[y \ll 1]{} [R_+/c\,|k|^{1/2}]\frac{2}{3}y^3 \qquad (D.16)$$

where $R_+/c\,|k|^{1/2} = 2.15 \times 10^{16}$ s, being $c = 3 \times 10^{10}$ cm/s and $|k|^{1/2} = \left(\frac{1}{2}\right)^{1/2} = 0.707$.

As shown in Chapter 7 observational data on $H_0 = \dot{R}_0/R_0$ and t_0 result in a value $y_0 = 0.243 \pm 0.157$ which implies a total cosmic mass:

$$M_u = R_+ c/2G\,|k|^{1/2} = 1.54 \times 10^{54}\,\text{g} \qquad (D.17)$$

very large but finite.

The cosmic expansion velocity as a function of the dimensionless cosmological parameter $y \equiv \sinh^{-1}(R/R_+)^{1/2}$ is:

$$y \ll 1 \quad \dot{R}(y) = [|k|^{1/2}c]/\tanh y \xrightarrow[y \ll 1]{} |k|^{1/2}c/y = (0.707)c/y \gg c \qquad (D.18)$$

Heisenberg's principle for the whole cosmos can be witten, assuming $\Delta R_{HL} \simeq R_{HL}$, $\Delta \dot{R}_{HL} \simeq \dot{R}_{HL}$ and $\Delta t_{HL} \simeq t_{HL}$, as

$$M_u \dot{R}_{HL} \cdot t_{HL} \simeq \hbar \qquad (D.19)$$

and as

$$M_u \dot{R}_{HL} \cdot R_{HL} \simeq \hbar \qquad (D.20)$$

where $R_{HL} = R(y_{HL})$, $\dot{R}_{HL} = \dot{R}(y_{HL})$, $t_{HL} = t(y_{HL})$, being y_{HL} the Heisenberg–Lemaitre value for the cosmologic dimensionless parameter y associated to the matter/energy mass $M_u/M_u c^2$.

Dividing Eq. (D.19) by Eq. (D.20) for $y_{HL} \ll 1$ we get

$$\frac{\dot{R}_{HL}}{R_{HL}} \cdot t_{HL} = H_{HL} \cdot t_{HL} \simeq 1 \qquad (D.21)$$

where the dimensionless product $H(y)t(y)$ for an open, finite universe is shown to be of the order **one**, as, in fact, was shown in Chapter 7, Eq. (7.15) for $y \ll 1$,

$$H(y) \cdot t(y) \xrightarrow[y \ll 1]{} \frac{2}{3} \qquad (D.22)$$

The thermal history of the open, finite universe can be followed going back from $t_0 = 13.7 \pm 0.2 \, \text{Gyrs} = 4.32 \times 10^{17} \, \text{s}$ to

$$t_{HL} = \hbar/M_u c^2 = 7.54 \times 10^{-103} \, \text{s}, \quad \text{much smaller than}$$

$$t_{Pl} = \hbar/m_{Pl} c^2 = 5.37 \times 10^{-44} \, \text{s} \quad \text{(Planck's time)}$$

To do this we must take into account that the **cosmic equation of state** changes at **decoupling/atom formation** from

$$RT = R_0 T_0 = (9.96 \times 10^{27} \, \text{cm})(2.726 \, {}^\circ\text{K}) \to y = \sinh^{-1}(R/R_+)^{1/2}$$

$$= \sinh^{-1}(T_+/T)^{1/2} > \to t > t_{\text{af}} \qquad (D.23)$$

$$R^3 T^4 = R_{\text{af}}^3 T_{\text{af}}^4 = (1.52 \times 10^{25} \, \text{cm})^3 (1.778 \times 10 \, {}^\circ\text{K})^4 \to$$

$$y = \sinh^{-1}(R/R_+)^{1/2} = \sinh^{-1}[(T_+/T_{\text{af}})(T_{\text{af}}/T)^{4/3}]^{1/2} \qquad (D.24)$$

$$< \to t < t_{\text{af}}$$

which allows one to evaluate $y(T)$ at primordial **nucleosyntheses** temperature, $T_{ns} = 4.6 \times 10^8\,°$K, **electron formation** temperature $T_e = m_e c^2 / 2.8 k_B$, **baryon formation** temperature $T_b = m_b c^2 / 2.8 k_B$ and beyond using $T_+ = 59.2\,°$K, and $T_{\mathrm{af}} = 1.77 \times 10^3\,°$K.

Heisenberg–Lemaitre time $t_{HL} \simeq 7.54 \times 10^{-103}$s corresponds to a cosmic temperature $T_{HL} \simeq M_u c^2 / 2.8 k_B = 3.58 \times 10^{90}\,°$K but Planck's time $t_{Pl} \simeq 5.36 \times 10^{-44}$s corresponds to a cosmic temperature $T_{Pl} \simeq m_{Pl} c^2 / 2.8 k_B = 5.05 \times 10^3\,°$K, very large but almost negligible in comparison with $T_{HL} = 3.58 \times 10^{90}\,°$K which is still physically meaningful, since it is not excluded by Heisenberg's **uncertainty** (not "**indeterminacy**") principle.

To conclude Appendix D, let us reproduce the main chronological events of these two great creative physicists of the twentieth century:

Werner Heisenberg (1901–1976)

1901 Werner Karl Heisenberg is born in Wurzburg (Germany) on 5th December.

1920 Enters the University of Munich and the Seminar of Arnold Sommerfeld.

1923 Ph. D. at the University of Munich, becomes assistant to Max and Born at the University of Gottingen.

1925 Heisenberg, Born and Jordan publish a seminal paper on the ground state and excited states of electrons in atoms and the quantum jumps between them through the absorption or emission of light.

1927 Heisenberg publishes his famous uncertainty principle.

1928 He is appointed professor of Theoretical Physics at the University of Leipzig.

1932 Sets forth his quantum model of the atomic nucleus in which neutron and proton can be viewed as two different quantum states of the same elementary particle.

1933 Receives the Nobel Prize in Physics 1932, for his contribution to set forth the formal basis of Quantum Mechanics.

1937 Contracts matrimony in Berlin with Elisabeth Schummacher.

1939 At the beginning of World War II becomes involved in the German nuclear project.

1942 Is named director of the Institute of Physics Kaiser Wilhelm in Berlin.

1943 Is appointed professor of Physics at the University of Berlin.

1945 At the end of World War II the Allies bring him to Farm Hall (England) in July 1945.

1946 He becomes director of the Max Planck Institute of Physics and Astrophysics I Gottingen.

1951 He is named president of the German Atomic Energy Commission and of the German delegation for the constitution of the CERN.

1976 Dies of cancer at his home in Munich on 1st February.

Georges Lemaitre (1894–1966)

1894 Georges Lemaitre is born in Belgium on 17 of July.

1914 Begins studying civil engineering at the Catholic University of Louvain. Interrupts his studies to serve as Artillery officer in World War I.

1918 Begins preparation for priesthood.

1920 Obtains his doctorate with a thesis entitled "L'Aproximation des functions de plusiers variables reales..." "under Professor" Ch. De la Vallee Poussin.

1923 Is ordained catholic priest. Spends a year at the University of Cambridge, UK working with Arthur Eddington on Cosmology, Stellar Astronomy and Numerical Analysis.

1924 Spends a year at Harvard College Observatory, Cambridge, Massachusetts with Harlow Shapley, and at MIT.

1925 Becomes part time lecturer at the Catholic University of Louvain, Belgium.

1927 Publishes his work "Un universe homogene de masse constant et de rayon croissant rendant compte de la vitesse radiale des nebulenses extragalactiques" where he derived Hubble's Law.

1931 Lemaitre translates into English his article with the help of Arthur Eddington but the part estimating the value of Hubble's "constant" is not translated. At this time Einstein tells Lemaitre "your calculations are correct but your physics is abominable".

He obtains his Ph. D. at MIT and becomes ordinary professor at the Catholic University of Louvain. He is invited to lecture in London at the meeting of the British Association and publishes his report on the "Primeval Atom in Nature. Later Fred Hoyle strongly critizices Lemaitre's theory calling it the "Big Bang Theory" and defends the "Steady State Theory", which postulates continuous creation of matter, supported by himself, Bondim, Gold.

1933 Lemaitre and Einstein travel together to California to give a series of seminars there.

1934 Lemaitre receives the highest Belgian decoration form King Leopold. A. Einstein, Ch. De la Vallee Poussin and A. Hemline are his scientific sponsors.

1936 He is elected member of the Pontificia Academy of Sciences.

1951 He disagrees privately with a famous discourse of Pope Pious XII before the Pontificia Academy of Sciences on the Proofs of God's existence in the light of Natural Modern Science.

1960 He is named president of the Academy during the Pontificate of Pope John XXIII.

1964 He suffers a heavy heart attack.

1966 He died on 20th of June shortly after having news of the discovery of Penzias and Wilson reporting the discovery of the Cosmic Background Radiation clearly compatible with the Big Band theory and very difficult to reconcile with the Steady State theory.

References

1. Julio A. Gonzalo (Coordinator). *Planck's Constants: 1900–2000* (Servicio de Publicaciones: Universidad Autónoma de Madrid, 2000), p. 103.
2. Julio A. Gonzalo, *Inflationary Cosmology Revisited* (World Scientific, Singapore, 2005).
3. Alan Guth, *The Inflationary Universe* (Perseus Books: Cambridge, Massachussetts, 1997).
4. H. Bondi, *Cosmology* (Cambridge University Press: Cambridge, 1952), p. 144.
5. R.A. Alpher and R. Herman, *Physics Today* **41** (Part 1) 1988.
6. A.A. Penzias and R. Wilson, *Astrophysical J.* **142**, 419 (1965).

7.   Julio A. Gonzalo, Jose L. Sánchez Gómez, and Miguel A. Alario (eds), *Cosmología Astrofísica* (Alianza Universidad: Madrid, 1996).

8.   Noé Cereceda, Ginés Lifante and Julio A. Gonzalo, "Acta Cosmologica" (Cracow) Fascicukus XXIV-2 (1998).

9.   Werner Heisenberg, "Zeitschrift für Physik" **43** (3–4): 172–198 (1927).

10.   Georges Lemaitre, "Annales de la Société de Bruxelles" **47**, 49 (April 1927).

Appendix E

The Medieval Roots of
Contemporary Science[*]

1. Why not in China?

Some enlightened European freethinkers contemporary of Voltaire, having read extensive reports on China by Jesuit missionaries, were very favourably impressed with the Chinese excellence in ethics and moral philosophy as well as in the arts and the crafts but found them lacking in science.[1] The two-volume work of Father Louis Lecompte, SJ, mathematician to his Majesty Louis XIV of France, entitled "Nouveaux Mémoires sur l'état present de la Chine" (1696) was composed of fourteen long letters to various French dignitaries, civic and ecclesiastical, went through five editions in five years, and was quickly translated into English, German and Dutch. Well written, with numerous illustrations, and printed in a handy format, the book covered many topics on Chinese geography, politics, history, literature arts, crafts and also science. The science that Father Lecompte had in mind was Euclid's and Ptolemys[2] but not the kind of science which, with roots in medieval Christendom, through Copernicus, Galileo and Kepler, had just achieved full maturity in the "Principia of Newton".

In view of the considerable talents, speculative and practical, evidenced in the millenary Chinese culture, those enlightened

---

[*]Articles below are based upon a lecture given by the author in Oviedo (Spain) on February 6th, 2007.

Europeans, considering the meagre scientific level achieved there asked themselves: Why no science in China?

In other words, why the astronomy, the mathematics and, especially, the physics in China was not comparable to the physical science in Europe at the time. The question raised was that of the historical origins of science, taking for science not a set of scattered partial achievements, more or less interesting, but not an organic system already in progress, as it was, already, European science by the closing years of the seventeen century. An organic system apparently destined to follow non-stop ascendant much towards what we know today as contemporary science.

But, as shown by Stanley L. Jaki in his work "Science and Creation: From Eternal Cycles to an Oscillating Universe",[3] science in this sense, had not had a viable birth neither in one of the great civilizations of Antiquity nor in Babylon, nor in Egypt, nor in India, not even in the most brilliant of them, the Classical Greek civilization. The Greeks had had spectacular developments in geometry and notable achievements in astronomy, but not in physics. In physics, i.e. in the study of mater in motion, and of the interactions between material bodies, they had made practically no progress. Their scientific knowledge in mechanics, optics, thermodynamics was rudimentary, if any. And the creative period, scientifically speaking, among the Greeks, had been relatively short. Greek Science came soon to a stand still, and was followed by a clear decline.

The question to be asked by those European freethinkers should have been another one: Why a genuine birth of science, a development susceptible of arriving, after some trial and error, to a final stage of self sustained scientific growth, had taken place in one civilization (our Western Christian Civilization) and only in it?

The question of how it began to develop in learned circles a firm and comprehensive scientific view of nature and its laws is certainly a momentous question, deserving a careful investigation and a balanced and objective response.

The earliest developments took place centuries ago in medieval Europe in a cultural matrix unequivocally Christian. There had been many other great civilizations in world history which could be proud of stupendous achievements in architecture, in public works, in dramatic arts, in ceramics, even in philosophy and logic. But in science proper, in physical science, no one had achieved, as already said, anything comparable to the achievements of our Western Christian Civilization.

The characteristic note of modern western science is its impressive capability to describe quantitatively a panoramic variety[4] of natural phenomena, from the realm of the elementary particles, to the evolution with time of the entire universe, through mathematical relations and differential equations, in a most precise and effective fashion. Nothing comparable had taken place in any of the great historical civilization, now extinct or alive. In our contemporary Western Civilization, with roots in medieval Christendom generations of scientists have made scientific progress following in their work a middle road between too much empiricism and too much idealism. Regardless of their ideological bent, the creative Western physicists, chemists, cosmologists of the last three centuries, when doing their work, did it as realists. Only in a realist perspective is it possible to escape the pitfalls which early scientific developments in other civilizations found in their way.

In those civilizations, the universe was either chaotic and incomprehensible or subject to iron deterministic laws of eternal returns. In either case there was no point in dedicating strenuous and sustained efforts, generation after generation, to understand it. In medieval Christian Europe, on the other hand, the world, created by God from nothing and in time, was basically good. It was well done, and man's intellect well made by God in his image and likeness, and well suited to perceive the unity, the truth and the beauty in the world.

If only matter is real and subject to purely chaotic behavior or to strictly deterministic laws, regardless of what the observer does or does not, everything will go on in its own chaotic or deterministic course. No need therefore for science. If, on the contrary, only ideas count, and the material world is radically false or perverse, there would be no point in the investigation of the natural laws which govern such a universe. The adventure of modern science has required indeed the efforts of many common men, and a handful of geniuses, through many generations. And it has been possible only through faith in the objective value of a universe well done, and confidence in man's intellectual ability to understand it.

If the world had not been rationally made, in other words, if something rationally valid one day were not valid the next day, or if a thing rationally valid here were invalid in any other more or less close neighborhood, systematic knowledge would be impossible.

It does not require to be a scientist to see the truth of these general considerations. The astonishing edifice of contemporary science, however, is a monumental proof of the fact that it is possible to describe quantitatively and systematically the laws governing nature, from the elementary particles to the most distant galaxies, and should be, therefore, a clear pointer to the wisdom of the Creator of nature and of the laws which govern it, as well as to the fact that man is endowed with intelligence and freedom. A freedom which allows him to recognize his Creator or to reject Him freely.

There is a subtle connection between a general theory of the physical world, a theory of knowledge and a natural theology, all three grounded in a Christian realism, heir of a revealed biblical wisdom which had come from the East and had infused a new life in the Western Greco-Roman Ecoumene at a time when the Roman Empire began to show signs of decline.

1.1. *Early medieval "natural philosophers"*

Abelard of Bath (fl. 1125) was one of those eager early medievals who went on long and arduous journeys in the quest of learning. He travelled as far as Aleppo, in the Middle East. Through him medieval Europe got access to trigonometry, to the description of astrolabs and to Euclid's geometry. His contacts with Muslim learned men made him aware of the struggle in the Arab culture to reconcile faith and reason. Having in mind his acquaintances with medieval Muslim and Jewish learned men, Abelard is on record remarking to his nephew that many of his contemporaries identified God with Nature. At his time, which was intellectually a turbulent time (but not only at his time) there was a strong tendency in men to identify nature with the ultimate entity, a tendency to be overcome only if men were willing to recognize humbly their dependence on a truly transcendent Creator. It is characteristic of Abelard in his "Questiones naturale"[5] the affirmation of the prerogatives of the Creation as well as those of the Creator. He favoured natural explanations over miraculous ones, whenever possible, showing his proclivity to a true scientific attitude facing the physical world: "I do not detract from God ... Whatever there is, is from Him and through Him. But the realm of being is not a confused one ... Only when reason totally fails, should the explanation of the matter be referred (directly) to God".

For medieval men as for men at anytime, the experience of personal freedom goes hand in hand with the experience of innate laziness. They too was tempted by the mirage of fatalism, which lured him towards astrological pantheism.

The biblical account of Creation, played a purifying role in the writings of Thierry of Chartres (d.c. 1155) rising him well above Greek animism and pantheism, even in its best literary expressions, such as Plato's "Timeus". "The intention of Moses (says Thierry) was to show that the creation of things and the formation of men was by the only one God, to whom alone worship is due. The utility of the work [of Moses] is the acquisition of knowledge about God through His handiwork". He saw Aristotle's four causes in a new perspective:

"Of the substances composing the world there are four causes: efficient, or God; formal, or God's wisdom; final, or His kindness; material, the four elements..." Note[6] that for Aristotle, as for the medievals, "Earth" included all solids, "water" meant all liquids, "air" all gases, and "fire" all heat. According to him all "elements", composing the world, "should have Him as their Creator, because they are all subject to change and they perish". For Thierry of Chartres, the speed of the moving heavens was much like the flight of a projectile: "when a stone is thrown, its impetus is ultimately due to the hold of the thrower against something solid [the ground]; the more firm is the stand of the one who throws, the more impetuous is the throw". For Thierry "there are four kinds of reasons that lead man to the recognition of his Creator: the proofs are taken from arithmetic, music [harmony], geometry and astronomy". If the Creator had actually arranged everything "according to number, measure, and weight", as recorded in the Book of Wisdom, man's understanding of the world had to reflect a mathematical character.

Robert Grosseteste (c. 1168–1253), Bishop of Lincoln, and possibly the first chancellor of the University of Oxford, a most influential figure in medieval thought, was even more explicit about the mathematical character of man's understanding of nature:

"The usefulness of considering lines, angles and figures is the greatest, because it is impossible to understand natural philosophy without them. They are efficacious throughout the universe as a whole and its parts, and in related properties as in rectilinear and linear motion". For instance, Grosseteste's investigation of the rainbow gives an excellent example of his scientific methodology, including seminal programs for induction, falsification, and verification. He, rightly, attributed the rainbows phenomenon to the refraction of light, rather than to its reflection, as done by Aristotle and Seneca. He considered all measurements made by man as imperfect, based as they were on "conventional units" and uncapable of counting the infinitely small, f.i. the (infinitely small) points contained in a line:

> "... since these are infinite, therefore their number cannot be known by a creature but by God alone, who disposes everything in number, weight and measure". According to S. L. Jaki, his still unedited treatises show that Grosseteste's scientific methodology depended on the idea of the Creator as wholly rational, personal Planner, Builder, and Maintainer of the Universe.

Not all the writings of the natural philosophers of the thirteen century were, of course, as clear and rational as Grosseteste's. Many were a mixture of insights, rumours, sound principles and fantastic tales, a combination of reason and magic. But even in some of the most confusing cases one can perceive the miracle of an age which "did not wholly succumb to the magical and irrational" in the quest for understanding, "unlike all previous cultures". For instance, in "De universo" of William of Auvergne (d.c. 1249), bishop of Paris, one can find frequent references to magical and astrological aphorisms, but can find also a continued polemics "against Manicheism, fatalism, pantheism, star worship" and similar betrayals of man's rationality and ultimate destiny, identified by him with the Saracens (Arabs), the Greeks of old and the Hermetic philosophers.

The lengthy discussion in "De universo" of the Great Year, identified with the 26,000 years period completing the precession of the equinoxes, shows its author awareness of the fact that belief of "eternal returns" every Great Year represented for William the farthest possible embodiment of the pagan, non-Christian world view.

Defending right reason William, aiming at the contemporary Mutakallimun, those devout learned Muslim philosophers, says: "Similarly, you must also part with those who in such matters take refuge in almighty God's most imperious will, and wholly abandon these questions as insoluble, and feel themselves at ease when they say that the Creator willed it that way, or that His will is the sole cause of all such things", and he points to the fatal error of not distinguishing between primary and secondary causality.

2. Tomas Aquinas and Roger Bacon

2.1. *Tomas Aquinas and the ways to God*

About the same time the great Saint Thomas Aquinas (1225–1274) was embarked in a gigantic effort to bring together faith and reason "in a stable synthesis". Moderate realism characterized his theory of knowledge, and the analogy of being formed the essence of his metaphysics. His resolute commitment to give reason his due meant as much as possible generous acceptance of the Aristotelian system, for two millennia the epitome of a rational explanation of the world. This commitment was motivated, as noted by Stanley Jaki, by the contemporary predicament of Muslim theologians and philosophers.

Aquinas completed his "Summa contra gentiles" (1257) with the aim of countering the occasionalism and the fatalism at that time contending with one another within Muslim contemporary culture. The work centered on questions about the Creator and about the nature of human intellect, trying always to depart little from Aristotle, except when the Christian Creed allowed no room for an agreement.

In his "Summa Theologica" (1273), a work of synthesis, not polemics, Aquinas went to a surprising extent in accepting Aristotle's cosmology and physics. Even to counteract the idea of an infinite endurance for the world, he resorted only to theological arguments. For Aquinas, the ultimate "raison dêtre" of the cosmos consisted in its subordination to man's supernatural destiny. Contrary to Aquinas, his master, Albertus Magnus was an enthusiastic advocate of experimental investigation, who found in the contingency of the world the justification of his prolific work on natural history. There was no difference, of course, between the disciple and the master, on the all important doctrine rejecting the inevitability of "eternal recurrences" in the world. But Albertus was more explicit in his dissertation on "De fato", about the history of the question of "eternal returns" in Plato, Aristotle, the Stoics, Ptolemy, the Arabs (especially Albumasar) partisans all of them of the "eternal returns", as well as in the early Church Fathers, who, of course, opposed it urgently.

In his "Summa Theologica" Saint Thomas raised the question: Whether God exists? Then he first presents two momentous objections. Objection 1: the existence of evil; Objection 2: the explanation of everything by natural causes, which makes God superfluous. Then he proceeds with his famous Five Ways,[7] beginning, all of them with specific characteristics of actual, existing beings.

First: From motion (change: generation/corruption).
      Arriving to a Prime Mover
Second: From causality.
      Arriving to an Uncaused Cause.
Third: From contingency.
      Arriving to a Necessary Being.
Fourth: From the degrees of perfection.
      Arriving to a Most Perfect Being.
Fifth: From order.
      Arriving to an Ordering Intellect.

As Saint Thomas says, this Prime Mover, Uncaused Cause, Necessary Being, Most Perfect Being, Ordering Intellect is what everyone understands to be God.

The most elegant and most metaphysical of the five is probably the Third Way: We find in nature contingent things (things which are possible to be or not to be). But it is impossible for these things always to exists, for that which is contingent cannot exist forever. Therefore, if everything were contingent, then, at one time there would be nothing in existence, because everything that exists begun to exist by something already existing. Which is absurd. Therefore the chain of contingent beings points necessary to a Necessary Being (Neither an infinite Chain made up only of contingent beings nor a closed circle of contingent beings explains the actual observed existence of contingent beings).

Take a star, for example, our Sun, a second or third generation star. We know today, Saint Thomas did not know (because at his time there were no NASA's artificial satellites, no Hubble's Telescopes, etc.), that an average mass star has a finite lifetime, perhaps ten thousand million years (longer if it is less massive, larger if it is more massive). This time is the time needed to consume all the hydrogen in its interior to produce helium and heavier elements. Then it will go through a phase as red giant, followed by another phase as white dwarf, to end up as a residual black dwarf. In case of stars more massive than the sun the resulting cosmic dust will become latter a newborn star, if physical conditions are propitious, under the gravitational attraction. And so forth... but not infinitely. The universe will be becoming more diluted and colder with the pass of time in a scale of tens of thousands of years. Stars are then contingent beings. And there is a finite number of generations of stars going back in time towards the Big Bang. We can trace back the matter and the energy to the time when atoms were first formed, out of the primeval plasma of electrons, protons and alpha particles. Then we can proceed backwards to the time when the elementary particles were formed in the presence of a very high radiation density. At these times no galaxies, no stars. And then we will approach times closer and closer to the Big Bang, till the uncertainty principle does not allow us to follow the time evolution of energy and material particles precisely.

The universe, therefore, is made up of contingent components, under the laws of physics (gravitation, radiation pressure, electromagnetic, strong and weak interactions ...) In all appearance the constituents of the universe as described by contemporary physics look contingent. The conservation of matter and energy (in ever more diluted form) the laws of physics, and the universal constants, all attest the cosmic stability and rationality. But Saint Thomas still is right: the actual existence of contingent beings requires the existence of a Necessary Being. A Necessary Being who is also, to begin with, a Prime Mover, an Uncaused Cause, a Most Perfect Being and an Ordering Intellect: "The wisdom of God (is) manifested in the works of Creation" as confirmed by the Book of Nature and the Book of Revelation.

2.2. *Roger Bacon and the experimental method*

Another influential man of the thirteen century was the franciscan Roger Bacon (1214–1294), medieval pioneer of the experimental method in the study of nature. Bacon's impetuous efforts to secure the service of science for the Christian faith lead him to compose his "Opus majus" (1267) which resulted in his temporary imprisionement, because some of his novel views stressing the inexorable determinism of events in nature appeared to his superiors as suspicious of incompatibility with man's freedom and moral responsibility. As noted by Stanley Jaki[2] his case shows clearly, nevertheless, "the difference between the Capitulation of most Arab commentators of Aristotle to the idea of cyclic determinism and Bacon's refusal to compromise in this crucial issue". He warned about the difference between final and efficient (secondary) causes, between man's supernatural destiny and his temporal (every day) subsistence. He asserted the forever partial character of man's knowledge about the world, in contrast to Aristotle's strict claims about *a priori*, definitive truths concerning the processes of nature.

In his "Opus majus" Bacon went as far as including sections on the application of astrology to Church government. Also, he was most emphatic on the existence of cycles in human history, paralleling the recurrence of planetary close encounters. But in the end he rejected the possibility of "eternal returns" of the world, because he saw them as incompatible with a Creator doing freely his work of creation. Taking the precession of the celestial axis (the period which he put as twenty-six thousand years) for evidence of the Great Year, would have been like following in the footsteps of Siger of Brabant (fl. 1260–1277). This would have been going too far in the advocacy of recurring the autonomy of philosophy from theology, as done by Averroes, the most distinguished Arab commentator of Aristotle. For Averroes the world was, as it was for Aristotle, a necessary emanation of the First Cause. In his grandiose, pantheistic, organismic universe of things, equivalent to a supreme animal, every event was believed to come back again and again with inexorable circularity. Aquinas, who was not a polemicist by nature, came forward immediately to discredit the "murmurantes" (Siger and his followers) in a public lecture on the subject of the World's eternity.

3. From Buridan and Oresme to Copernicus and Newton

In 1277 Etienne Tempier, bishop of Paris condemned 219 propositions covering a wide spectrum of questions clearly excluding the eternity of the world (Prop. 83–91) and the perennial recurrence of everything every 26,000 years (Prop. 92). In substance the bishop of Paris was affirming rigorously the contingency of the world with respect to a transcendental Creator, source of all rationality in the macrocosm as well as in the microcosm, in heaven as well as in Earth.

The decree bishop Tempier was taken by Pierre Duhem, the great French physicist and pioneer historian of physics, as the starting point of a new era in scientific thinking.

In the decree it was recognized the possibility of several worlds (something which perhaps could attract some sympathy from some present day partisans of extra-terrestrial intelligent beings) (Prop. 27); the rejection of superlunary bodies as animated, incorruptible and eternal (Prop. 31–32); the possibility of rectilinear motion for celestial bodies as part of a celestial machinery (Prop. 75); the rejection of a deterministic influence of celestial stars on the life of individual men from the instant of birth (Prop. 105); the rejection of the provenance of the "first matter" from celestial matter (Prop. 107); all of it aimed at defending the exclusive rights and freedom of the Creator when he created. This attitude shaped the conceptual framework for a revolutionary new approach towards the understanding of the motion of celestial as well as terrestrial bodies. The statement of Paris' bishop was not binding on the universal Church, but it was primarily a university affair, brought up by the fact that Siger of Brabante had been teaching for over a decade at the University of Paris, at which bishop Tempier was charged with the duty of maintaining orthodoxy.

In his "Lowell Lectures" (1925) A. N. Whitehead[9] said: "I do not think however, that I have even yet brought up the greatest contribution to the formation of the scientific movement. I mean the inexpugnable belief that every detailed occurrence can be correlated with its antecedents in a perfectly definite manner, exemplifying general principles (emphasis added). Without this belief the incredible labours of scientists would be without hope. It is this instructive conviction, vividly posed before the imagination, which is the motive power of research: that there is a secret, a secret which can be unveiled. How has this conviction been so vividly implanted in the European mind?"

There was an intimate connection between the rational conjectures of medieval "physicists" and their firm believes in a personal Creator. This can be seen in their comments on "De Caelo" (On the Heavens), the systematic exposition of Aristotle's world views. Of extraordinary importance in this respect is John Buridan's "Questioness super quattor libris de caelo et mundo"[10] through its slightly modified version by his disciple Albert of Saxony, which, according to Stanley Jaki had year after an unmistakable influence on Galileo. Copies of this work were commonly available at medieval European universities, from Salamanca to Krakow.

The Aristotelian clear-cut distinction between superlunary and sublunary matter was dealt a decisive blow by Buridan. While Aristotle denied that the heaven could decay, Buridan reminded his readers that they could decay and even that the Creator was perfectly capable of annihilating the world.[11]

Buridan warned against confusing faith and reason, and offered an explanation of the plurality observed in the heavenly motions which did justice to both theology and natural science.

He noted that "in natural philosophy one should consider process and causal relationships as if they always come about in a natural fashion; God is no less the cause, therefore, if this world and its order have an end than if this world was eternal".

Actually he did not yet accept as a fact the rotation of the Earth on its axis (his discipline Oresme would do it latter) but, against Aristotle, he maintained the continuous though relatively small changes of the Earth respect the fixed stars, which, for him, implied that the Earth was not at the center of the universe. This and other considerations represented an enormous step in the right direction, away from the ancient pagan worldview epitomized by Aristotle.

Buridan, against the mistaken notions on motion as due to continuous contact of the mover with the moved body, and on the idea of the natural attraction of a body to its "natural" place, proposed his notion of "impetus", a quality implanted in the moving body by its mover, and the notion of "gravity", a property innate to all massive bodies. And, after reviewing the usefulness of his theory with respect to various movements down here on Earth, he further dared to outline its usefulness for celestial mechanics. In the same breath he wrote of jump, of planetary motion, and of the Creator as the ultimate agent imparting a given quantity of motion to the various parts of the universe:

"[If one] Whishes to jump a long distance [goes] back a way in order to run faster, so that... he might acquire an impetus which would carry him a longer distance in the jump... the person so running and jumping does not feel the air [as] moving him, but [rather] feels the air in front Strongly Resisting him.

Also, since the Bible does not say that appropriate intelligences move the celestial bodies, it could be said that it does so appear necessary to posit intelligences of this kind... [rather it could be said that] God, when He created the world, moved each of the celestial orbs as He pleased, and in moving them He impressed in them impetuses which moved them without His having to move them anymore except by the... general influence whereby He concurs as a co-agent in all things which take place... And these impetuses which He impressed in the celestial bodies were not decreased not

corrupted afterwards, because there was no inclination [to it]. . .
Nor was there resistance which would be corruptive or resistive
of that impetus. . . But His I do not say assertively, but [rather
tentatively]. . . "[12]

Nicole Oresme (1323–1382) was the most outstanding disciple of
Burindan. At the request of Charles the V, king of France, Oresme
undertook the translation and interpretation of the three major works
of Aristotle, the "Nicomean Ethics", the "Politics" and "On the
Heaven". While he was at the College of Navarre, of which he became
grand master in 1536, he did original work on a wide range of topics
including monetary theory, astronomy, geometry and algebra. In his
"De configurationibus qualitatum et motuum" he was trying to get a
universal mathematical method to describe physical changes as wall
as changes in man's inner experiences, psychological and esthetic.
His commentary to "On the Heaven" is considered today as a great
classic of scientific literature. Oresme's work has received encomiastic
support from great pioneer historians of science such as P. Duhem
and H. Dingler, but has received a more critical appraisal from other
historians such as A. Koyré, one of the most influential positivist historians of science in the last century. Of course there is good justification in recognizing the newness of Galilean Science, but the primitive
steps in the same direction in the writings of Burindan and Oresme
deserve as much justified recognition, as a fact of history, with an
enormous impact on the modern mind, proud of its science.

With reference to Aristotle's insistence on the perfection of the
universe, Oresme said that the perfection of the laws of nature were
but a modest reflection of the infinitely perfect attributes of the
Creator. Oresme allowed incorruptibility to the celestial bodies and
the celestial motions only in the restricted sense that these motions
were frictionless, allowing a unified perspective to discuss earthly
and heavenly motions. Against the possibility of eternal recurrences
he noted in his "De proportionibus proportionum" that the periods
of the planets are most likely incommensurable and, therefore, that
they cannot return exactly to the same relative position. In any case,
he noted, such periods should be much larger than the 26,000 years
of the period of precession of the equinoxes, the length assigned in

ancient Greece to the "Great Year" after which everything would be repeated again. Belief in the "Great Year" had yielded to a new era of thinking, rooted in the liberating influence of the doctrine about creation by an absolutely sovereign Creator. Oresme even speculated on the possibility of worlds enclosed within worlds. Oresme parted with Aristotelian necessitarianism implying that several words would require several Gods. He responded: "One God would govern all such worlds"[13]

According to Aristotle the Prime Mover's perfection imposed a strictly spherical shape on the world. Oresme considered this an unreasonable restriction on the Creators power, admitting the possibility of empty space either outside or inside the stellar realm. It was a very remarkable anticipation of the section in Newton's Scholium to his "Principia" in which the idea of an infinite empty space is conceptually linked to the infinity of the Creator. Oresme said: "... Outside the heavens, there is an empty incorporeal space quite different from any other plenum or corporeal space, just as the extent of this time called eternity is of a different sort than temporal duration, even if the latter were perpetual... Now this space of which we are talking is infinite and indivisible, and is the immensity of God and God himself. Just as the duration of God called eternity is infinite, indivisible, and God himself, as already stated above"[14]

Clearly, for Oresme, God was transcendent to the world, and his transcendence could be safeguarded by viewing him as imparting a given quantity of motion (impetus) to the world once and for all.

Later "impetus" would be correctly redefined as "momentum", i.e. inertial mass times the velocity imparted to the moving body. This crucial conceptual development was impossible to conceive within the pantheistic necessitarianism of Aristotle.

But the full process from Buridam and Oresme to Newton took three hundred years. Copernicus (1473–1543) with the impetus theory behind him, gave the crucial step of postulating the heliocentric description of the inertial planetary movements, with references to how the Psalmist said properly that God was immensely pleased by his handywork. Then Galileo (1564–1642) championed vigorously the Copernican system, and did so nothing that according to the Church

Fathers (Saint Augustine in particular) the Bible said "how to go to Heaven, not how the Heavens go". It may be pointed out also that his arguments based on the tides to support the Copernican system were flawed and that two hundred years would pass before the first evidence of the parallax of a star was observationally confirmed.

Paris' Cathedral.

And Johannes Kepler (1571–1630), luckily named assistant to Tycho Brahe in Prague's observatory inherited Tycho's precise observational records of the planetary evolutions, leading to the formulation of his three famous laws: (1) The planetary orbits are ellipses; (2) The areas swept by the vector radii in equal times are equal; and (3) The ratio between the cubes of the semi-major axis of the ellipses and their corresponding squared period were the same for all solar planets then known. Everything was ready for Isaac Newton (1642–1727) to formulate precisely the laws of classical mechanics and the law of universal gravitation noting the analogy between the fall of an apple and the slow fall of the moon on Earth.

Thus, the centuries of medieval European Christendom, which witnessed the creation of the medieval universities: Bologna, Paris, Oxford, Salamanca, Prague, Krakow..., and the impressive gothic cathedrals: Chartres, Paris, Oviedo, Leon, Burgos... with their unsurpassed elegance and beauty, were also the centuries in which the roots of the edifice of modern science begun to take hold in the conceptual developments adumbrated by Buridam, Oresme and their medieval predecessors.

References

1. S. L. Jaki, *The Origin of Science and the Science of its Origin* (Reguery/Gateway, Inc.: South Bend, Indian, 1978), pp. 22–42.
2. Fr. Louis Lecompte, *Nouveaux Memoires sur l'etat present de la Chine* (Chez Jean Anisson: Paris, 1696). [Quoted in Ref. 1]
3. S. L. Jaki, *Science and Creation: From Eternal Cycles to an Oscillating Universe* (University Press of America: Lanhan MD, 1990).
4. Julio A. Gonzalo, *Pioneros de la ciencia* (Palabra: Madrid, 2000).
5. S. L. Jaki, *Ibid* (Ref. 3), p. 219.
6. Virginia Trimble, FHP Newsletter, APS, Vol. X, No. 1, Fall 2006 p. 6.
7. Peter Kreeft, *Summa of the Summa* (Ignatius Press: San Francisco, 1990).
8. S. L. Jaki, *Science and Creation: From Eternal Cycles to an Oscillating Universe* (University Press of America: Lanhan MD, 1990), p. 227.
9. A. N. Whitehead, *Science and the Modern World: Lowell Lectures, 1925* (The Macmillan Company: New York, 1925).
10. Edited by E. A. Moody (Cambridge, Mass: The Medieval Academy of America, 1942).
11. *Ibid.*, p. 90 (Lib. I, quaest. 14).
12. M. Clagett, *The Science of Mechanics in the Middle Ages* (University of Wisconsin Press: Madison, 1959), p. 536.
13. N. Oresme, *Le Livre du ciel et du monde*, p. 171 (Book I, Chap. xxiv).
14. *Ibid.*, p. 177.

Glossary

Alpha particle. Nucleus of ^{4}He atom made up of two neutrons and two protons. After primeval nucleosynthesis the cosmic plasma is made up mainly by protons (76%), α particles (24%) and electrons.

Baryon-to-photon ratio. At present the universe is *transparent* and the ratio of baryons to photons in the universe is $n_b/n_r \approx 10^{-9}$ and is kept constant. It is to be expected that before atom formation, at temperatures above 3 000 to 4 000 K, the universe was in an *opaque* plasma state, in which there was considerable friction (multiple scattering) and the density of photons increased with time with respect to the density of baryons (nuclei of H and ^{4}He) (See Chapter 12).

Baryogenesis. Process by which a definite population of baryons (protons, neutrons) becomes well defined in the early universe. This requires that the cosmic matter density becomes equal or less than nuclear density, which occurs at a cosmic temperature $T_b \approx 3.88 \times 10^{12}$ K. It must have taken place at $t \sim 10^{-5}$ s.

Big Bang nucleosynthesis. Primordial process by which cosmic protons and neutrons fuse together to form ^{4}He nuclei and traces of other light nuclei. This occurs at a temperature $T_{ns} \sim 4.60 \times 10^8$ K. It must have taken place at $t \sim 10^3$ s.

Blueshift. If a star or galaxy moves toward us the radiation emitted by it observed from the Earth appears shifted toward shorter wavelength (towards the blue) (See Doppler shift).

Cepheid variable. A type of variable star whose brightness was seen to vary over time. Henrietta Swann Leavitt discovered in 1912 that an approximately linear relationship existed between the period of a Cepheid's pulsation and its luminosity. The longer the period of its pulsation, the greater was the star's luminosity. Once the distance to nearby Cepheids was determined, they became useful measuring sticks for establishing the distances to other galaxies after they were discovered. In 1925 Edwin Hubble discovered a Cepheid in M31, the Andromeda Nebula, which enabled him to estimate that it was 8×10^5 light-years from Earth. It convinced astronomers that Andromeda was not part of the Milky Way, but a galaxy in its own right.

Cosmic accelerated expansion. Recent observations of magnitude (related to distance) versus redshift (related to recession velocity) which indicate that the rate of change of velocity with distance is larger for nearby galaxies than for far-away galaxies. To properly interpret the evolution in rate of change it is necessary to take into account that the maximum recession velocity cannot exceed the speed of light and that for the most distant protogalaxies relativistic effects must be expected (See Chapter 10).

Cosmic time parameter $H \times t$. Cosmic parameter (dimensionless) giving the product of the Hubble's parameter $H = \dot{R}/R$ and the cosmic time at any moment in the cosmic expansion. For an open universe $H \times t$ is less than one and more than two third $(2/3)$. This parameter is time dependent, because $H \times t$ is time dependent and, of course, t is also time dependent. At present, using local data, $H \cdot t \approx 0.94 \pm 0.06$, substantially larger than $H_{af} t_{af} \approx 0.667 \approx 2/3$, corresponding to the time of atom formation $(t_{af} \approx 300.000 \, \text{yrs})$.

Dark energy. Since $\Omega \approx 1$ is consistent with the estimated matter-energy density at the time of "decoupling" $(T \approx 3\,000 \, \text{K})$, and the dark mass has been taken as contributing about 30% to it, the difference (about 70%) is commonly attributed to the repulsive force of the vacuum (cosmological constant) in the form of a kind of potential "dark energy": For a discussion of the time dependence of the matter mass density $(\Omega_m(t))$ and the potential space-time curvature energy density $(\Omega_k(t))$ resulting in $1 = \Omega_m(t) + \Omega_k(t)$, compatible with

$\langle \Omega_m \rangle \approx 0.26$, $\langle \Omega_k \rangle \approx 0.74$, averaged from the time of galaxy formation to the present, see Chapter 11.

Dark matter. In the close neighborhood of our galaxy the matter mass density in galaxies is only about 4% of the critical density (required for galaxies being exactly at *escape* velocities). It is commonly assumed that in order to have $\Omega = \rho/\rho_c = 1$. Over cosmic space-time a matter mass density of at least 30% is required, including matter of non-baryonic nature. This is the so called "dark matter". In discussing dark matter, the time dependence of $\Omega(t)$ from very early times to present as given by the Friedmann–Lemaitre solutions of Einstein's cosmological equations, is usually ignored (see Chapter 11).

Dark night paradox. The paradox which points out the apparent impossibility of having a dark night in a universe with an infinite number of luminous stars homogeneously distributed.

Doppler shift. Shift in the receiver frequency (and wavelength) of waves (sound or electromagnetic radiation) due to the relative motion of source and observer, depending on their approach or recession. For light, i.e., electromagnetic radiation, results in a blueshift or a redshift, respectively. Christian Doppler first proposed this effect in terms of the change in pitch for sound waves in 1841, but modern astronomers apply it to electromagnetic waves, corresponding to an observed change in color. Stars or galaxies with spectra shifted to the blue (or shorter) wavelengths of the spectrum are moving toward the Earth. Stars and nebulae with spectra shifted to the red, or longer, wavelengths are moving away.

Electromagnetic interactions. Interactions due to electric charges. Static charges give rise to electric fields, accelerated/decelerated charges produce electromagnetic waves, including radio, microwave, infrared, visible, ultraviolet light and X-rays, as well as γ-rays.

Electroweak interactions. Unified description of weak interactions (responsible for nuclear β decay) and electromagnetic interactions, due to Glashow, Weinberg and Abdus Salam (1967–1970).

Energy conservation. Formulated by nineteenth century physicists (originally by Julius R. Mayer in 1842). It is the principle that establishes the strict *conservation* of energy: energy is not destroyed or created in physical processes, it is only transformed, for instance, thermal energy into mechanical energy, or vice versa. After the advent of Relativity Theory, associating an energy $E = mc^2$ to any mass m, the principle was generalized to include the energy associated to rest mass.

Expansion of the universe. The expansion of cosmic space suggested by Einstein's field equations of general relativity. Alexander Friedmann showed in 1922 that such an expansion was a natural consequence of the field equations, but Einstein, at first disagreed with Friedmann, who died in 1925.

Finiteness of the universe. We assume the universe to be intelligible (otherwise we would not be investigating it). It is *infinite* or *finite*. If it is infinite it is beyond our capability of understanding. So we *conclude* it is finite, and hope not to incur a contradiction, as pointed out explicitly by Georges Lemaitre, and as required explicitly by Einstein.

Hubble parameter. The ratio of recession velocity of galaxies ($\dot{R}$) to distance (R), improperly called Hubble's constant, because for nearby galaxies $H_o \approx \dot{R}_o/R_o$ is approximately constant, but, in principle, it is time-dependent, being $H(t) \approx \dot{R}(t)/R(t) \gg H(t_o)$ for early times. The presently accepted value of the Hubble parameter is $H_o \sim 67(\text{km/s})/\text{Mpc}$, with 1 Mpc (Megaparsec) $= 3.26 \times 10^6$ light years.

Hubble time. Estimate for the age of the universe based on the reciprocal of Hubble's parameter. At the time Lemaitre originally proposed his primeval atom theory, the Hubble time was estimated to be only about two billion years. Later revisions of Hubble's distance estimates increased the value upward to 4×10^9 years in 1948 and upward to 1×10^{10} by the time Allan Sandage took over Hubble's position at Mount Palomar.

Isotropy. Attribute assumed by Lemaitre, and other cosmologists, that the universe appears the same in all directions. To a very high degree, measurements of the cosmic background radiation roughly confirms this.

Leptons. A class of non-strongly interacting particles which includes the electron, the muon, the tau particle, and their associates.

Magnetic monopole. An isolated pole (north or south) of a hypothetic magnet. An infinitely long magnet in which the two poles are so far apart that they do not affect each other would act as a pair of monopoles. Dirac predicted the existence of particles with properties of magnetic monopoles, and so did some grand unified theories. But they have never been observed.

Photon. A quantum (discrete) minimum of electromagnetic energy consisting essentially in a localized particle of light after the quantum theory of radiation put forward by Max Planck in 1900.

Quarks. Elementary constituent of protons, neutrons and other strongly interacting particles (baryons and mesons).

Red giant. A phase in the life cycle of a typical star (not very heavy, like our Sun) in which the size of the star increases tremendously and then blows up into space, leaving as a remnant a white dwarf.

Steady state theory. A cosmological theory postulated by Fred Hoyle, Thomas Gold, and Herman Bondi in Britain in 1948, asserting that the universe is in essence without change. Hoyle, Bondi, and Gold argued against a temporal, Big Bang origin, suggesting that while the universe was expanding, the expansion occurred in a natural state similar to that suggested by de Sitter's universe, but with matter being constantly created in the form of hydrogen atoms in empty space, keeping cosmic conditions identical as a function of time. In the early 1960s the discovery of quasars and other developments in astronomy strengthened the contention that the universe in the distant past was far different from the universe at present. The discovery of the cosmic microwave background radiation all but discredited the state the theory.

Type 1a supernovae. Violent stellar explosions signaling the utter demise of a star, that take place when a white dwarf bleeds enough mass from a companion star to pass a certain mass limit, approximately 1.4 times the mass of the Sun (also called Chandrasekhar's limit, after the Indian-American astrophysicist Subrahmanyan Chandrasekhar), at which point it explodes. The luminance of such supernovae is so great that they can be brighter than entire galaxies as they are photographed. The consistent level of luminosity of Type 1a supernovae have made them an even more reliable distance indicator than Cepheids for determining extragalactic distances.

White dwarf. Final phase in the life of a typical star after it went through the phase of red giant.

Author Index

Abraham, Isaac, Jacob, 115
Alfonseca, M., 69, 76
Alonso, M., 43
Alpher, R. A., vii, xvi, 24, 33, 53, 57, 80, 81, 84, 128
Astier, P., 90

Bentley, R., 42
Bessel, F. W., 127
Bohr, N., 10, 21, 63, 66, 107, 108
Boltzmann, L., 65, 66
Bolyai, J., 38, 39, 127
Boslough, J., 52
Born, M., 65, 67
Bondi, H., vii, 11, 13, 28, 32
Brawer, R., 53, 57
Buridan, 118

Calvin, J., 127
Cannon, R. D., 84
Cantarell, I., 12
Carroll, S., 84
Cereceda, N., 95
Chaboyer, B., 90
Chandrasekhar, S., 166
Chernin, A., 62
Cheseaux, J. P. L., 27, 32
Compton, A. H., 21, 22, 103, 104
Copernicus, N., 119
Cowan, C., 10
Crespo, E., 63
Crick, F., 113, 116

Dalí, S., 114
de Broglie, L., 22, 107, 108
Debye, P., 22, 63
Denton, M., 116
Dicke, R. H., 19, 27
Dirac, P., 19

Eddington, A. S., 5, 33, 53, 60
Einstein, 39
Einstein, A., 5, 6, 9, 10, 22, 25, 28, 29, 32, 33, 35, 37–39, 41, 42, 45, 47, 49, 53, 59–63, 65, 69–71, 73, 74, 89, 104–106, 108, 109, 128
Engels, F., 87
Eisberg, R., 19, 25
Elcano, J. S., 127
Evans, R. D., 95

Farrell, J., 8, 62
Finn, E. J., 43
Friedmann, A., 45
Fritzsch, H., 19

Gárate, A., 63
Galileo, G., 127
Gamow, G., vii, 24, 52, 60
Gauss, K. F., 38, 39, 105, 127
Godart, O., 8
Gold, T., vii, 11, 28, 32
Gonzalez, G., 19, 116
Gonzalo, J. A., xvi, 8, 13, 19, 25, 52, 67, 76, 95

Subject Index

Printed in the United States
by Baker & Taylor Publisher Services